Springer Series in Reliability Engineering

Series Editor

Professor Hoang Pham
Department of Industrial Engineering
Rutgers
The State University of New Jersey
96 Frelinghuysen Road
Piscataway, NJ 08854-8018
USA

Other titles in this series

The Universal Generating Function in Reliability Analysis and Optimization
Gregory Levitin

Warranty Management and Product Manufacture
D.N.P Murthy and Wallace R. Blischke

Maintenance Theory of Reliability
Toshio Nakagawa

Reliability and Optimal Maintenance
Hongzhou Wang and Hoang Pham

System Software Reliability
Hoang Pham

B.S. Dhillon

Applied Reliability and Quality

Fundamentals, Methods and Procedures

Springer

B. S. Dhillon, PhD
Department of Mechanical Engineering
University of Ottawa
Ottawa, Ontario
Canada

British Library Cataloguing in Publication Data
Dhillon, B. S. (Balbir S.), 1947–
Applied reliability and quality : fundamentals, methods and applications. – (Springer series in reliability engineering)
1. Reliability (Engineering) 2. Quality control
I.Title
620'.00452
ISBN-13: 9781846284977
ISBN-10: 184628497X

Library of Congress Control Number: 2006939314

Springer Series in Reliability Engineering series ISSN 1614-7839
ISBN 978-1-84628-497-7 e-ISBN 978-1-84628-498-4 Printed on acid-free paper

9 8 7 6 5 4 3 2 1

Springer Science+Business Media
springer.com

This book is affectionately dedicated
to my colleague Professor Stavros Tavoularis
for helping me to trace my ancient Scythian ancestry
that resulted in the publication of a book on the matter
and for challenging me to write 30 books.

Foreword

In today's technological world nearly everyone depends upon the continued functioning of a wide array of complex machinery and equipment for our everyday safety, security, mobility, and economic welfare. We expect our electric appliances, hospital monitoring control, next-generation aircraft, data exchange systems, banking, and aerospace applications to function whenever we need them. When they fail, the results can be catastrophic. As our society grows in complexity, there is a need to understand the critical reliability challenges. In other words, people want to know how reliable of their products by understanding how to quantify the reliability and quality of existing systems.

This volume, **Applied Reliability and Quality**, which is a well-written introduction in 11 chapters, is designed for an introductory course on applied reliability and quality for engineering and science students as well as a suitable for a short training course on applied engineering reliability. The book consists of four parts. The first part discusses some fundamental elements of probability and statistics including probability properties, basic statistics measures, and some common distribution functions. The second part presents various introductory aspects of reliability engineering and their applications in medical devices, electrical power, and robotic and computer systems. It also includes various methods such as Markov, fault tree, and failure modes and effect analysis. The third part describes some fundamental concepts of quality control and assurance and its applications in healthcare, software engineering, textiles, and the food industry. Finally, the last part provides a comprehensive list of literature references for readers who are interested in obtaining additional information on this subject.

Each chapter provides a basic introduction to applied engineering reliability and quality, an unusually diverse selection of examples, and a variety of exercises designed to help the readers further understand the material.

Hoang Pham
Series Editor

Preface

Today, billions of dollars are being spent annually world wide to develop reliable and good quality products and services. Global competition and other factors are forcing manufacturers and others to produce highly reliable and good-quality products and services. Needless to say, nowadays reliability and quality principles are being applied across many diverse sectors of economy and each of these sectors has tailored reliability and quality principles, methods, and procedures to satisfy its specific need. Some examples of these sectors are robotics, health care, electric power generation, Internet, textile, food, and software.

There is a definite need for reliability and quality professionals working in diverse areas to know about each other's work activities because this may help them, directly or indirectly, to perform their tasks more effectively. At present to the best of author's knowledge, there is no book that covers both applied reliability and quality within its framework. More specifically, at present to gain knowledge of each other's specialties, these specialists must study various books, articles, or reports on each of the areas in question. This approach is time consuming and rather difficult because of the specialized nature of the material involved.

This book is an attempt to meet the need for a single volume that considers applied areas of both reliability and quality. The material covered is treated in such a manner that the reader needs no previous knowledge to understand it. The sources of most of the material presented are given in the reference section at the end of each chapter. At appropriate places, the book contains examples along with solutions, and at the end of each chapter there are numerous problems to test reader comprehension. This will allow the volume to be used as a text. A comprehensive list of references on various aspects of applied reliability and quality is provided at the end of this book, to give readers a view of the intensity of developments in the area.

The book is composed of 11 chapters. Chapter 1 presents need for applied reliability and quality, reliability and quality history, important reliability and quality terms and definitions, and sources for obtaining useful information on applied reliability and quality. Chapter 2 reviews various mathematical concepts considered useful to understand subsequent chapters. Some of these concepts are arithmetic mean, mean deviation, standard deviation, Laplace transform definition,

Newton method, Boolean algebra laws and probability properties, and probability distributions. Chapter 3 presents various introductory aspects of both reliability and quality.

Chapter 4 is devoted to robot reliability. It covers topics such as robot failure causes and classifications, robot reliability measures, robot reliability analysis methods, and models for performing robot reliability and maintenance studies. Chapters 5 presents medical equipment reliability-related topics such as medical equipment reliability improvement procedures and methods, human error in medical equipment, guidelines for reliability and other professionals to improve medical equipment reliability, and organizations and sources for obtaining medical equipment failure-related data.

Chapter 6 is devoted to power system reliability and covers topics such as service performance indices, loss of load probability, models for performing availability analysis of a single generator unit, and models for performing availability analysis of transmission and associated systems. Chapter 7 presents various aspects of computer and Internet reliability including computer system failure causes and measures, fault masking, software reliability evaluation models, Internet failure examples and outage categories, and Internet reliability models. Chapters 8 and 9 are devoted to quality in health care and software quality, respectively.

Chapter 10 covers various important aspects of quality control in the textile industry including quality-related issues in textiles, textile quality control department functions, textile test methods, and quality control in spinning and fabric manufacture. Chapter 11 is devoted to quality control in the food industry. It covers topics such as factors affecting food quality, basic elements of a food quality assurance program, the hazard analysis and critical control points (HACCP) concept, fruits and vegetables quality, and food processing industry quality guidelines.

This book will be useful to many people including design engineers, manufacturing engineers, system engineers, engineering and manufacturing managers, reliability specialists, quality specialists, graduate and senior undergraduate students of engineering, researchers and instructors of reliability and quality, and professionals in areas such as health care, software, electric power generation, robotics, textile, food, and the Internet.

The author is deeply indebted to many individuals including colleagues, friends, and students for their invisible inputs and encouragement throughout the project. I thank my children Jasmine and Mark for their patience and invisible inputs. Last, but not least, I thank my other half, friend, and wife, Rosy, for typing various portions of this book and other related materials, and for her timely help in proofreading.

Ottawa, Ontario B.S. Dhillon

Contents

1

Introduction

1.1 Need for Applied Reliability and Quality

Today, billions of dollars are being spent annually worldwide to develop reliable and good quality products and services. Global competition and other factors are forcing manufacturers and others to produce highly reliable and good quality products and services. Needless to say, reliability and quality principles are being applied across many diverse sectors of the economy, and each of these sectors has tailored reliability/quality principles, methods, and procedures to satisfy its specific need. Some examples of these sectors are robotics, health care, electric power generation, the Internet, textile, food, and software.

As a result, there is a definite need for reliability and quality professionals working in diverse areas such as these to know about each others' work activities because this may help them, directly or indirectly, to perform their tasks more effectively. In turn, this will result in better reliability and quality of end products and services.

1.2 Reliability and Quality History

This section presents an overview of historical developments in both reliability and quality areas, separately.

1.2.1 Reliability History

The history of the reliability field may be traced back to the early 1930s, when probability principles were applied to electric power generation-related problems in the United States [1–5]. During World War II, Germany applied the basic reliability concepts to improve reliability of their V1 and V2 rockets. Also during World War II, the United States Department of Defense recognized the need for reliability improvement of its equipment. During the period between 1945–1950, it performed various studies concerning the failure of electronic equipment, equipment maintenance and repair cost, etc. The results of three of these studies were as follows [6]:

- An Army study indicated that between two-thirds and three-fourths of the equipment used by the Army was either out of commission or under repair.
- An Air Force study performed over a period of five years revealed that repair and maintenance costs of the equipment used by the Air Force were approximately ten times the original cost.
- A Navy study conducted during maneuvers revealed that the electronic equipment used was functional about 30% of the time.

As the result of studies such as these, the US Department of Defense established an ad hoc committee on reliability, in 1950. In 1952, this committee became a permanent group known as the Advisory Group on the Reliability of Electronic Equipment (AGREE). In 1957, the Group released its report, called the AGREE report, that ultimately resulted in the release of a specification on the reliability of military electronic equipment [6].

In 1954, a National Symposium on Reliability and Quality Control was held for the first time in the United States. Two years later, in 1956, the first commercially available book on reliability was published [7]. The first master's degree program in system reliability engineering was started at the Air Force Institute of Technology of the United States Air Force (USAF) in 1962.

All in all, ever since the inception of the reliability field many people and organizations have contributed to it and a vast number of publications on the subject have appeared [8, 9]. A more detailed history of the developments in the reliability field is available in [8, 10].

1.2.2 Quality History

The history of the quality field may be traced back to the ancient times to the construction of pyramids by the ancient Egyptians (1315–1090 BC). During their construction quality-related principles were followed, particularly in regard to workmanship, product size, and materials. In the 12th century AD quality standards were established by the guilds [11].

However, in the modern times (*i. e.*, by 1907) the Western Electric Company was the first to use basic quality control principles in design, manufacturing, and installation. In 1916, C.N. Frazee of Telephone Laboratories successfully applied statistical approaches to inspection-related problems, and in 1917, G.S. Radford coined the term "quality control" [12]. In 1924, Walter A. Shewhart of Western Electric Company developed quality control charts. More specifically, he wrote a memorandum on May 16, 1924, that contained a sketch of modern quality control chart. Seven years later, in 1931, he published a book entitled *Economic Control of Quality of Manufactured Product* [13].

In 1944, the journal *Industrial Quality Control* was jointly published by the University of Buffalo and the Buffalo Chapter of the Society of Quality Control Engineers. In 1946, the American Society for Quality Control (ASQC) was formed, and this journal became its official voice.

Over the years, many people and organizations have contributed to the field of quality and a vast number of publications on the topic have appeared [8, 14]. A large number of publications on four applied areas of quality are listed at the end of this book. A more detailed history of the developments in the field of quality is available in [8, 11, 14, 15].

1.3 Reliability and Quality Terms and Definitions

There are a large number of terms and definitions currently being used in reliability and quality areas. Some of the commonly used terms and definitions in both these areas are presented below, separately [16–22].

1.3.1 Reliability

- **Reliability.** This is the probability that an item will perform its stated mission satisfactorily for the specified time period when used under the stated conditions.
- **Failure.** This is the inability of an item to function within the stated guidelines.
- **Hazard rate (instantaneous failure rate).** This is the rate of change of the number of items that have failed over the number of items that have survived at time t.
- **Availability.** This is the probability that the equipment is operating satisfactorily at time t when used according to specified conditions, where the total time considered includes active repair time, operating time, logistic time, and administrative time.
- **Redundancy.** This is the existence of more than one means to accomplish a stated function.
- **Maintainability.** This is the probability that a failed item will be restored to its satisfactory working state.
- **Reliability engineering.** This is the science of including those factors in the basic design that will assure the specified degree of reliability, maintainability, and availability.
- **Reliability demonstration.** This is evaluating equipment/item capability to meet stated reliability by actually operating it.
- **Reliability model.** This is a model to predict, assess, or estimate reliability.
- **Reliability growth.** This is the improvement in a reliability figure-of-merit caused by successful learning or rectification of faults in equipment/item design, manufacture, sales, service, or use.

1.3.2 Quality

- **Quality.** This is the degree to which an item, function, or process satisfies the needs of users and customers.
- **Quality control.** This is a management function, whereby control of raw materials' and manufactured items' quality is exercised to stop the production of defective items.
- **Quality plan.** This is the documented set of procedures that covers the in-process and final inspection of product.
- **Quality control program.** This is an overall structure that serves to define the quality control system objectives.
- **Quality control engineering.** This is an engineering approach whereby technological skills and experiences are utilized to predict quality attainable with various designs, production processes, and operating set-ups.
- **Control chart.** This is the chart that contains control limits.
- **Random sample.** This is a sample of units in which each unit has been chosen at random from the source lot.
- **Sampling plan.** This is a plan that states the sample size to be inspected and provides acceptance and rejection numbers.
- **Process inspection.** This is intermittent examination and measurement with emphasis on the checking of process variables.
- **Quality assurance.** This is a planned and systematic sequence of all actions appropriate for providing satisfactory confidence that the product/item conforms to established technical requirements.
- **Quality measure.** This is a quantitative measure of the features and characteristics of an item or service.
- **Quality management.** This is the totality of functions involved in achieving and determining quality.
- **Average incoming quality.** This is the average level of quality going into the inspection point.

1.4 Useful Information on Applied Reliability and Quality

There are many sources for obtaining applied reliability and quality-related information. Some of the most useful sources for obtaining such information, on both reliability and quality, are presented below, separately, under many different categories [8, 9, 23].

1.4.1 Reliability

Organizations

- Reliability Society, IEEE, P.O. Box 1331, Piscataway, New Jersey, U.S.A.
- SOLE, the International Society of Logistics, 8100 Professional Place, Suite 111, Hyattsville, Maryland, U.S.A.
- American Society for Quality Control, 310 West Wisconsin Avenue, Milwaukee, Wisconsin, U.S.A.
- National Aeronautics and Space Administration (NASA) Parts Reliability Information Center, George C. Marshall Space Flight Center, Huntsville, Alabama, U.S.A.
- Reliability Analysis Center, Rome Air Development Center, Griffis Air Force Base, Rome, New York, U.S.A.
- System Reliability Service, Safety and Reliability Directorate, UKAEA, Wigshaw Lane, Culcheth, Warrington, U.K.

Journals

- *Reliability Engineering and System Safety*
- *International Journal of Quality and Reliability Management*
- *IEEE Transactions on Reliability*
- *Engineering Failure Analysis*
- *Quality and Reliability Engineering International*
- *Software Testing, Verification, and Reliability*
- *Risk Analysis*
- *Microelectronics Reliability*
- *Journal of Machinery Manufacture and Reliability*
- *Journal of Risk and Reliability*
- *Reliability Review*
- *International Journal of Reliability, Quality, and Safety Engineering*
- *Journal of the Reliability Analysis Center*
- *Lifetime Data Analysis*
- *Reliability Magazine*
- *IEEE Transactions on Power Apparatus and Systems*

Conference Proceedings

- *Proceedings of the Annual Reliability and Maintainability Symposium*
- *Proceedings of the ISSAT International Conference on Reliability and Quality in Design*

- *Proceedings of the Annual International Reliability, Availability, and Maintainability Conference for the Electric Power Industry*
- *Proceedings of the European Conference on Safety and Reliability*

Books

- *Grant Ireson, W., Coombs, C.F., Moss, R.Y., Editors, Handbook of Reliability Engineering Management, McGraw Hill Book Company, New York, 1996.*
- *Dhillon, B.S., Reliability Engineering in Systems Design and Operation, Van Nostrand Reinhold Company, New York, 1983.*
- *Ohring, M., Reliability and Failure of Electronic Materials and Devices, Academic Press, San Diego, California, 1998.*
- *Kumar, U.D., Crocker, J., Chitra, T., Reliability and Six Sigma, Springer, New York, 2006.*
- *Thomas, M.U., Reliability and Warranties: Methods for Product Development and Quality Improvement, Taylor and Francis, Boca Raton, Florida, 2006.*
- *Billinton, R., Allan, R.N., Reliability Evaluation of Power Systems, Plenum Press, New York, 1996.*
- *Dhillon, B.S., Design Reliability: Fundamentals and Applications, CRC Press, Boca Raton, Florida, 1999.*
- *Shooman, M.L., Probabilistic Reliability: An Engineering Approach, McGraw Hill Book Company, New York, 1968.*
- *Dhillon, B.S., Medical Device Reliability and Associated Areas, CRC Press, Boca Raton, Florida, 2000.*

Standards and Other Documents

- MIL-STD-785B, Reliability Program for Systems and Equipment, Development, and Production, US Department of Defense, Washington, D.C.
- MIL-STD-721C, Definition of Terms for Reliability and Maintainability, US Department of Defense, Washington, D.C.
- MIL-HDBK-338, Electronic Reliability Design Handbook, US Department of Defense, Washington, D.C.
- MIL-HDBK-217F, Reliability Prediction of Electronic Equipment, US Department of Defense, Washington, D.C.
- MIL-STD-790E, Reliability Assurance Program for Electronic Parts Specifications, US Department of Defense, Washington, D.C.
- MIL-STD-1629A, Procedures for Performing a Failure Mode, Effects, and Criticality Analysis, US Department of Defense, Washington, D.C.
- MIL-STD-2155, Failure Reporting, Analysis and Corrective Action System (FRACAS), US Department of Defense, Washington, D.C.
- MIL-HDBK-189, Reliability Growth Management, US Department of Defense, Washington, D.C.

- MIL-HDBK-781, Reliability Test Methods, Plans, and Environments for Engineering Development, Qualification, and Production, US Department of Defense, Washington, D.C.
- MIL-STD-781D, Reliability Design Qualification and Production Acceptance Tests: Exponential Distribution, US Department of Defense, Washington, D.C.
- MIL-STD-756, Reliability Modeling and Prediction, US Department of Defense, Washington, D.C.

1.4.2 Quality

Organizations

- American Society for Quality Control, 310 West Wisconsin Avenue, Milwaukee, Wisconsin, U.S.A.
- European Organization for Quality, 3 rue du Luxembourg, B-1000, Brussels, Belgium.
- American Society for Testing and Materials, 1916 Race Street, Philadelphia, Pennsylvania, U.S.A.
- American National Standards Institute (ANSI), 11 W. 42nd St., New York, New York, U.S.A.
- Government Industry Data Exchange Program (GIDEP), GIDEP Operations Center, U.S. Department of Navy, Naval Weapons Station, Seal Beach, Corona, California, U.S.A.
- National Technical Information Service (NTIS), 5285 Port Royal Road, Springfield, Virginia, U.S.A.

Journals

- *Quality Progress*
- *Quality in Manufacturing*
- *Benchmarking for Quality Management and Technology*
- *International Journal of Quality and Reliability Management*
- *Journal of Quality in Maintenance Engineering*
- *Journal of Quality Technology*
- *Quality Forum*
- *Quality Today*
- *International Journal of Health Care Quality Assurance*
- *Managing Service Quality*
- *Quality Assurance in Education*
- *The TQM Magazine*
- *International Journal for Quality in Health Care*
- *Quality Engineering*

- *Six Sigma Forum Magazine*
- *Software Quality Professional*
- *Techno-metrics*
- *Quality Management Journal*
- *Journal for Quality and Participation*
- *The Quality Circle Journal*
- *Quality Assurance*
- *Industrial Quality Control*
- *Quality Review*

Conference Proceedings

- *Transactions of the American Society for Quality Control (Conference Proceedings)*
- *Proceedings of the European Organization for Quality, Conferences*
- *Proceedings of the Institute of Quality Assurance Conferences (U.K.)*

Books

- *Beckford, J., Quality, Routledge, New York, 2002.*
- *McCormick, K., Quality, Butterworth Heinemann, Boston, 2002.*
- *Kirk, R., Healthcare Quality and Productivity: Practical Management Tools, Aspen Publishers, Rockville, Maryland, 1988.*
- *Alli, I., Food Quality Assurance: Principles and Practices, CRC Press, Boca Raton, Florida, 2004.*
- *Schroder, M.J.A., Food Quality and Consumer Value: Delivering food that satisfies, Springer, Berlin, 2003.*
- *Vardeman, S., Jobe, J.M., Statistical Quality Assurance Methods for Engineers, John Wiley and Sons, New York, 1999.*
- *Gryna, F.M., Quality Planning and Analysis, McGraw Hill Book Company, New York, 2001.*
- *Rayan, T.P., Statistical Methods for Quality Improvement, John Wiley and Sons, New York, 2000.*
- *Shaw, P., et al., Quality and Performance Improvement in Healthcare: A Tool for Programming Learning, American Health Information Management Association, Chicago, 2003.*
- *Galin, D., Software Quality Assurance, Pearson Education Limited, New York, 2004.*
- *Meyerhoff, D., editor, Software Quality and Software Testing in Internet Times, Springer, New York, 2002.*
- *Kemp, K.W., The Efficient Use of Quality Control Data, Oxford University Press, New York, 2001.*

- *Hartman, MG., Editor, Fundamentals Concepts of Quality Improvement, ASQ Quality Press, Milwaukee, Wisconsin, 2002.*
- *Smith, G.M., Statistical Process Control and Quality Improvement, Prentice Hall, Inc., Upper Saddle River, New Jersey, 2001.*
- *Bentley, J.P., An Introduction to Reliability and Quality Engineering, John Wiley and Sons, New York, 1993.*
- *Kolarik, W.J., Creating Quality: Process Design for Results, McGraw Hill Book Company, New York, 1999.*
- *Evans, J.R., Lindsay, W.M., The Management and Control of Quality, West Publishing Company, New York, 1989.*

Standards and Other Documents

- *ANSI/ASQC A3, Quality Systems Terminology, American National Standards Institute (ANSI), New York.*
- *MIL-HDBK-53, Guide for Sampling Inspection, US Department of Defense, Washington, D.C.*
- *ANSI\ASQC B1, Guide for Quality Control, American National Standards Institute (ANSI), New York.*
- *MIL-STD-52779, Software Quality Assurance Program Requirements, US Department of Defense, Washington, D.C.*
- *ANSI/ASQC E2, Guide to Inspection Planning, American National Standards Institute (ANSI), New York.*
- *MIL-HDBK-344, Environmental Stress Screening of Electronic Equipment, US Department of Defense, Washington, D.C.*
- *MIL-STD-2164, Environment Stress Screening Process for Electronic Equipment, US Department of Defense, Washington, D.C.*
- *MIL-STD-105, Sampling Procedures and Tables for Inspection by Attributes, US Department of Defense, Washington, D.C.*
- *ANSI/ASQC A1, Definitions, Symbols, Formulas, and Table for Quality Charts, American National Standards Institute (ANSI), New York.*
- *ANSI/ASQC B2, Control Chart Method for Analyzing Data, American National Standards Institute (ANSI), New York.*
- *ANSI/ASQC A2, Terms, Symbols and Definitions for Acceptance Sampling, American Standards Institute (ANSI), New York.*

1.5 Problems

1. Define and compare reliability and quality.
2. List at least five areas of applied reliability.
3. Discuss historical developments in the area of quality.

4. Define the following terms:
   - Reliability growth
   - Hazard rate
   - Reliability engineering
5. What is the difference between quality control and quality control engineering?
6. Define the following quality-related terms:
   - Quality plan
   - Quality management
   - Process inspection
7. What is the difference between quality assurance and quality control?
8. List five important organizations for obtaining reliability-related information.
9. Write an essay on the history of the reliability field.
10. Discuss at least three important organizations for obtaining quality-related information.

References

1 Layman, W.J., Fundamental Consideration in Preparing a Master System Plan, Electrical World, Vol. 101, 1933, pp. 778–792.
2 Smith, S.A., Service Reliability Measured by Probabilities of Outage, Electrical World, Vol. 103, 1934, pp. 371–374.
3 Smith, S.A., Spare Capacity Fixed by Probability of Outage, Electrical World, Vol. 103, 1934, pp. 222–225.
4 Benner, P.E., The Use of the Theory of Probability to Determine Spare Capacity, General Electric Review, Vol. 37, 1934, pp. 345–348.
5 Smith, S.A., Probability Theory and Spare Equipment, Edison Electric Inst. Bull., March 1934, pp. 110–113.
6 Shooman, M.L., Probabilistic Reliability: An Engineering Approach, McGraw Hill Book Company, New York, 1968.
7 Henney, K., Lopatin, I., Zimmer, E.T., Adler, L.K., Naresky, J.J., Reliability Factors for Ground Electronic Equipment, McGraw Hill Book Company, New York, 1956.
8 Dhillon, B.S., Reliability and Quality Control: Bibliography on General and Specialized areas, Beta Publishers, Gloucester, Ontario, 1992.
9 Dhillon, B.S., Reliability Engineering Applications: Bibliography on important application areas, Beta Publishers, Gloucester, Ontario, 1992.
10 Coppola, A., Reliability Engineering of Electronic Equipment: A Historical Perspective, IEEE Transactions on Reliability, Vol. 33, 1984, pp. 29–35.
11 Hayes, G.E., Romig, H.G., Modern Quality control, Collier Macmillan Publishers, London, 1977.
12 Radford, G.S., Quality Control (Control of Quality), Industrial Management, Vol. 54, 1917, pp. 100.

13 Shewhart, W.A., Economic Control of Quality of Manufactured Product, D. Van Nostrand Company, New York, 1931.
14 Krismann, C., Quality Control: An Annotated Bibliography, The Kraus Organization Limited, White Plains, New York, 1990.
15 Golomski, W.A., Quality Control: History in the Making, Quality Progress, Vol. 9, No. 7, July 1976, pp. 16–18.
16 Omdahl, T.P., Editor, Reliability, Availability, and Maintainability (RAM) Dictionary, ASQC Quality Press, Milwaukee, Wisconsin, 1988.
17 MIL-STD-721, Definitions of Effectiveness Terms for Reliability, Maintainability, Human Factors, and Safety, Department of Defense, Washington, D.C.
18 Naresky, J.J., Reliability Definitions, IEEE Transactions on Reliability, Vol. 19, 1970, pp. 198–200.
19 Von Alven, W.H., Editor, Reliability Engineering, Prentice Hall, Inc., Englewood Cliffs, New Jersey, 1964.
20 Lester, R.H., Enrich, N.C., Mottley, H.E., Quality Control for Profit, Industrial Press, New York, 1977.
21 McKenna, T., Oliverson, R., Glossary of Reliability and Maintenance, Gulf Publishing Company, Houston, Texas, 1977.
22 ANSI/ASQC A3-1978, Quality Systems Terminology, American Society for Quality Control, Milwaukee, Wisconsin.
23 Dhillon, B.S., Design Reliability: Fundamentals and Techniques, CRC Press, Boca Raton, Florida, 1999.

2

Reliability and Quality Mathematics

2.1 Introduction

Since mathematics has played a pivotal role in the development of quality and reliability fields, it is essential to have a clear understanding of the mathematical concepts relevant to these two areas. Probability concepts are probably the most widely used mathematical concepts in both reliability and quality areas. The history of probability may be traced back to the sixteenth century, when a gambler's manual written by Girolamo Cardano (1501–1576) made reference to probability.

However, it was not until the seventeenth century when Pierre Fermat (1601–1665) and Blaise Pascal (1623–1662) solved the problem of dividing the winnings in a game of chance correctly and independently. Over the years many other people have contributed in the development of mathematical concepts used in the fields of reliability and quality. More detailed information on the history of mathematics and probability is available in [1,2]. More specifically, both these documents are totally devoted to the historical developments in mathematics and probability. This chapter presents mathematical concepts considered useful to understand subsequent chapters of this book.

2.2 Arithmetic Mean, Mean Deviation, and Standard Deviation

These three measures are presented below, separately.

2.2.1 Arithmetic Mean

This is expressed by

$$m = \frac{\sum_{i=1}^{n} DV_i}{n} \tag{2.1}$$

where
n is the number of data values.
DVi is the data value i; for i = 1, 2, 3, …, n.
m is the mean value (*i. e.*, arithmetic mean).

Example 2.1
The quality control department of an automobile manufacturing company inspected a sample of 5 identical vehicles and discovered 15, 4, 11, 8, and 12 defects in each of these vehicles. Calculate the mean number of defects per vehicle (*i. e.*, arithmetic mean).

Using the above-specified data values in Equation (2.1) yields

$$m = \frac{15+4+11+8+12}{5}$$
$$= 10 \textit{ defects per vehicle}$$

Thus, the mean number of defects per vehicle or the arithmetic mean of the data set is 10.

2.2.2 Mean Deviation

This is one of the most widely used measures of dispersion. More specifically, it is used to indicate the degree to which a given set of data tend to spread about the mean. Mean deviation is expressed by

$$m_d = \frac{\sum_{i=1}^{n} \left| DV_i - m \right|}{n} \tag{2.2}$$

where

n is the number of data values.

DVi is the data value i; for i = 1, 2, 3, …, n.

m_d is the mean value deviation.

m is the mean of the given data set.

$|DV_i - m|$ is the absolute value of the deviation of DVi from m.

Example 2.2

Find the mean deviation of the Example 2.1 data set.

In Example 2.1, the calculated mean value of the data set is 10 defects per vehicle. By substituting this calculated value and the given data into Equation (2.2), we get

$$m_d = \frac{[|15-10|+|4-10|+|11-10|+|8-10|+|12-10|]}{5}$$

$$= \frac{[5+6+1+2+2]}{5}$$

$$= 3.2$$

Thus, the mean deviation of the Example 2.1 data set is 3.2.

2.2.3 Standard Deviation

This is expressed by

$$\sigma = \left[\frac{\sum_{i=1}^{n} (DV_i - m)^2}{n} \right]^{1/2} \tag{2.3}$$

where

σ is the standard deviation.

Standard deviation is a commonly used measure of data dispersion in a given data set about the mean, and its three properties pertaining to the normal distribution are as follows [3]:

- 68.27% of the all data values are within $m + \sigma$ and $m - \sigma$.
- 95.45% of the all data values are within $m - 2\sigma$ and $m + 2\sigma$.
- 99.73% of the all data values are within $m - 3\sigma$ and $m + 3\sigma$.

Example 2.3
Find the standard deviation of the data set given in Example 2.1.

Using the calculated mean value, m, of the given data set of Example 2.1 and the given data in Equation (2.3) yields

$$\begin{aligned}\sigma &= \left[\frac{(15-10)^2+(4-10)^2+(11-10)^2+(8-10)^2+(12-10)^2}{5}\right]^{1/2}\\ &= \left[\frac{5^2+(-6)^2+1^2+(-2)^2+2^2}{5}\right]^{1/2}\\ &= 3.74\end{aligned}$$

Thus, the standard deviation of the data set given in Example 2.1 is 3.74.

2.3 Some Useful Mathematical Definitions and Formulas

There are many mathematical definitions and formulas used in quality and reliability fields. This section presents some of the commonly used definitions and formulas in both these areas.

2.3.1 Laplace Transform

The Laplace transform is defined by [4] as

$$F(s) = \int_0^\infty f(t)e^{-st}\,\mathrm{d}t \tag{2.4}$$

where
t is time.
s is the Laplace transform variable.
$F(s)$ is the Laplace transform of function, $f(t)$.

Laplace transforms of four commonly occurring functions in reliability and quality work are presented in Table 2.1. Laplace transforms of other functions can be found in [4, 5].

Table 2.1. Laplace transforms of four commonly occurring functions

$f(t)$	$F(s)$
$e^{-\theta t}$	$\frac{1}{s+\theta}$
$\frac{d f(t)}{dt}$	$s F(s) - f(0)$
c (constant)	$\frac{c}{s}$
t	$\frac{1}{s^2}$

2.3.2 Laplace Transform: Initial-Value Theorem

If the following limits exist, then the initial-value theorem may be stated as

$$\lim_{t \to 0} f(t) = \lim_{s \to \infty} s F(s) \tag{2.5}$$

2.3.3 Laplace Transform: Final-Value Theorem

Provided the following limits exist, then the final-value theorem may be stated as

$$\lim_{t \to \infty} f(t) = \lim_{s \to 0} s F(s) \tag{2.6}$$

2.3.4 Quadratic Equation

This is defined by

$$ay^2 + by + c = 0 \quad \textit{for } a \neq 0 \tag{2.7}$$

where
a, b, and c are constants.

Thus,

$$y = \frac{-b \pm \left(b^2 - 4ac\right)^{1/2}}{2a} \tag{2.8}$$

If a, b, and c are real and $M = b^2 - 4ac$ is the discriminant, then the roots of the equation are

- Real and equal if $M = 0$
- Complex conjugate if $M < 0$
- Real and unequal if $M > 0$

If y_1 and y_2 are the roots of Equation (2.7), then we can write the following expressions:

$$y_1 + y_2 = \frac{-b}{a} \tag{2.9}$$

and

$$y_1 \, y_2 = \frac{c}{a} \tag{2.10}$$

Example 2.4
Solve the following quadratic equation:

$$y^2 + 13y + 40 = 0 \tag{2.11}$$

Thus, in Equation (2.11), the values of a, b, and c are 1, 13, and 40, respectively. Using these values in Equation (2.8) yields

$$y = \frac{-13 \pm \left[13^2 - 4(1)(40)\right]^{1/2}}{2(1)}$$

$$y = \frac{-13 + 3}{2}$$

Therefore,

$$y_1 = \frac{-13 + 3}{2}$$
$$= -5$$

and

$$y_2 = \frac{-13 - 3}{2}$$
$$= -8$$

Thus, the roots of Equation (2.11) are $y_1 = -5$ and $y_2 = -8$. More specifically, both these values of y satisfy Equation (2.11).

2.3.5 Newton Method

Newton's method is a widely used method to approximate the real roots of an equation that involves obtaining successive approximations. The method uses the following formula to approximate real roots of an equation [6, 7]:

$$x_{n+1} = x_n - \frac{f(x_n)}{f'(x_n)}, \ for \ f'(x_n) \neq 0 \tag{2.12}$$

where
the prime (′) denotes differentiation with respect to x.
x_n is the value of the nth approximation.

The method is demonstrated through the following example.

Example 2.5
Approximate the real roots of the following equation by using the Newton's approach:

$$x^2 - 26 = 0 \tag{2.13}$$

As a first step, we write

$$f(x) = x^2 - 26 \tag{2.14}$$

By differentiating Equation (2.14) with respect to x, we get

$$\frac{\mathrm{d}f(x)}{\mathrm{d}x} = 2x \tag{2.15}$$

Inserting Equations (2.14) and (2.15) into Equation (2.12) yields

$$x_{n+1} = x_n - \frac{x_n^2 - 26}{2x_n} = \frac{x_n^2 + 26}{2x_n} \tag{2.16}$$

For $n = 1$ in Equation (2.16) we chose $x_1 = 5$ as the first approximation. Thus, Equation (2.16) yields

$$x_2 = \frac{x_1^2 + 26}{2x_1} = \frac{(5)^2 + 26}{2(5)} = 5.1$$

For $n = 2$, substituting the above-calculated value into Equation (2.16), we get

$$x_3 = \frac{x_2^2 + 26}{2x_2} = \frac{(5.1)^2 + 26}{2(5.1)} = 5.099$$

Similarly, for $n=3$, substituting the above-calculated value into Equation (2.16), we get

$$x_4 = \frac{x_3^2 + 26}{2x_3} = \frac{(5.099)^2 + 26}{2(5.099)} = 5.099$$

It is to be noted that the values x_3 and x_4 are the same, which simply means that the real root of Equation (2.13) is $x=5.099$. It can easily be verified by substituting this value into Equation (2.13).

2.4 Boolean Algebra Laws and Probability Properties

Boolean algebra is named after mathematician George Boole (1813–1864). Some of the Boolean algebra laws that can be useful in reliability and quality work are as follows [8, 9]:

$$A.B = B.A \tag{2.17}$$

where

A is an arbitrary event or set.

B is an arbitrary event or set.

Dot (.) between A and B or B and A denotes the intersection of events or sets. However, sometimes Equation (2.17) is written without the dot, but it still conveys the same meaning.

$$A + B = B + A \tag{2.18}$$

where

$+$ denotes the union of sets or events.

$$A(B + C) = AB + AC \tag{2.19}$$

where

C is an arbitrary set or event.

$$(A + B) + C = A + (B + C) \tag{2.20}$$

$$AA = A \tag{2.21}$$

$$A + A = A \tag{2.22}$$

$$A(A + B) = A \tag{2.23}$$

$$A + AB = A \tag{2.24}$$

$$(A + B)(A + C) = A + BC \tag{2.25}$$

As probability theory plays an important role in reliability and quality, some basic properties of probability are as follows [10–12]:

- The probability of occurrence of event, say X, is

$$0 \le P(X) \le 1 \tag{2.26}$$

- Probability of the sample space S is

$$P(S) = 1 \tag{2.27}$$

- Probability of the negation of the sample space is

$$P(\overline{S}) = 1 \tag{2.28}$$

where

$\overline{S}$ is the negation of the sample space S.

- The probability of occurrence and non-occurrence of an event, say X, is

$$P(X) + P(\overline{X}) = 1 \tag{2.29}$$

where

$P(X)$ is the probability of occurrence of event X.

$P(\overline{X})$ is the probability of non-occurrence of event X.

- The probability of an interaction of K independent events is

$$P(X_1 X_2 X_3 \ldots . X_K) = P(X_1)P(X_2)P(X_3) \ldots . P(X_K) \tag{2.30}$$

where

X_i is the ith event; for $i = 1, 2, 3, \ldots ., K$.

$P(X_i)$ is the probability of occurrence of event X_i; for $i = 1, 2, 3, \ldots ., K$.

- The probability of the union of K independent events is

$$P(X_1 + X_2 + X_3 + \ldots . + X_K) = 1 - \prod_{i=1}^{K} \left(1 - P(X_i)\right) \tag{2.31}$$

For $K = 2$, Equation (2.32) reduces to

$$P(X_1 + X_2) = P(X_1) + P(X_2) - P(X_1)P(X_2) \tag{2.32}$$

- The probability of the union of K mutually exclusive events is

$$P(X_1 + X_2 + X_3 + \ldots . + X_K) = P(X_1) + P(X_2) + P(X_3) + \ldots + P(X_K) \tag{2.33}$$

2.5 Probability-related Mathematical Definitions

There are various probability-related mathematical definitions used in performing reliability and quality analyses. Some of these are presented below [10–13]:

2.5.1 Definition of Probability

This is expressed by [11]

$$P(Y) = \lim_{m \to \infty} \left[\frac{M}{m} \right] \tag{2.34}$$

where

$P(Y)$ is the probability of occurrence of event Y.

M is the total number of times Y occurs in the m repeated experiments.

2.5.2 Cumulative Distribution Function

For a continuous random variable, this is expressed by

$$F(t) = \int_{-\infty}^{t} f(y)\,\mathrm{d}y \tag{2.35}$$

where

t is time (*i. e.*, a continuous random variable).

$F(t)$ is the cumulative distribution function.

$f(t)$ is the probability density function (in reliability work, it is known as the failure density function).

2.5.3 Probability Density Function

This is expressed by

$$f(t) = \frac{\mathrm{d}F(t)}{\mathrm{d}t} = \frac{\mathrm{d}\left[\int_{-\infty}^{t} f(y)\,\mathrm{d}y \right]}{\mathrm{d}t} \tag{2.36}$$

2.5.4 Expected Value

The expected value, $E(t)$, of a continuous random variable is expressed by

$$E(t) = M = \int_{-\infty}^{\infty} t\, f(t)\, \mathrm{d}t \tag{2.37}$$

where
$E(t)$ is the expected value of the continuous random variable t.
$f(t)$ is the probability density function.
M is the mean value.

2.5.5 Variance

This is defined by

$$\theta^2(t) = E(t^2) - [E(t)]^2 \tag{2.38}$$

or

$$\theta^2(t) = \int_{0}^{\infty} t^2 f(t)\, \mathrm{d}t - M^2 \tag{2.39}$$

where
$\sigma^2(t)$ is the variance of random variable t.

2.6 Statistical Distributions

In mathematical reliability and quality analyses, various types of probability or statistical distributions are used. Some of these distributions are presented below [13, 14].

2.6.1 Binomial Distribution

The binominal distribution is named after Jakob Bernoulli (1654–1705) and is used in situations where one is concerned with the probabilities of outcome such as the total number of occurrences (*e. g.*, failures) in a sequence of, say m, trials [1]. However, it should be noted that each trial has two possible outcomes (*e. g.*, success and failure), but the probability of each trial remains constant.

The distribution probability density function is defined by

$$f(x) = \binom{m}{x} p^x q^{m-x}, \; for \; x = 0, 1, 2, \ldots, m \tag{2.40}$$

where
$f(x)$ is the binomial distribution probability density function.

$$\binom{m}{x} = \frac{m!}{x!(m-x)!}$$

x is the number occurrences (*e. g.*, failures) in m trials.
p is the single trial probability of success.
$q = 1 - p,$ is the single trial probability of failure.

The cumulative distribution function is given by

$$F(x) = \sum_{i=0}^{x} \frac{m!}{i!(m-i)!} p^i q^{m-i} \tag{2.41}$$

where
$F(x)$ is the cumulative distribution function or the probability of x or less failures in m trials.

The mean or the expected value of the distribution is [10]

$$E(x) = m\,p \tag{2.42}$$

2.6.2 Poisson Distribution

The Poisson distribution is named after Simeon Poisson (1781–1840), a French mathematician, and is used in situations where one is interested in the occurrence of a number of events that are of the same type. More specifically, this distribution is used when the number of possible events is large, but the occurrence probability over a specified time period is small. Waiting lines and the occurrence of defects are two examples of such a situation. The distribution probability density function is defined by

$$f(x) = \frac{\lambda^x e^{-\lambda}}{x!}, \; for\ x = 0, 1, 2, \ldots \tag{2.43}$$

where
λ is the distribution parameter.

The cumulative Poisson distribution function is

$$F(x) = \sum_{i=0}^{x} \frac{\lambda^i e^{-\lambda}}{i!} \tag{2.44}$$

The mean or the expected value of the distribution is [10]

$$E(x) = \lambda \tag{2.45}$$

2.6.3 Normal Distribution

Although normal distribution was discovered by De Moivre in 1733, time to time it is also referred to as Gaussian distribution after German mathematician, Carl Friedrich Gauss (1777–1855). Nonetheless, it is one of the most widely used continuous random variable distributions, and its probability density function is defined by

$$f(t) = \frac{1}{\sigma\sqrt{2\Pi}} \exp\left[-\frac{(t-\mu)^2}{2\sigma^2}\right], -\infty \langle t \langle +\infty \tag{2.46}$$

where

t is the time variable.

μ and σ are the distribution parameters (*i. e.*, mean and standard deviation, respectively).

By substituting Equation (2.46) into Equation (2.35) we get the following equation for the cumulative distribution function:

$$F(t) = \frac{1}{\sigma\sqrt{2\Pi}} \int_{-\infty}^{t} \exp\left[-\frac{(x-\mu)^2}{2\sigma^2}\right] dx \tag{2.47}$$

Using Equation (2.46) in Equation (2.37) yields the following expression for the distribution mean:

$$E(t) = \mu \tag{2.48}$$

2.6.4 Gamma Distribution

The Gamma Distribution is a two-parameter distribution, and in 1961 it was considered as a possible model in life test problems [15]. The distribution probability density function is defined by

$$f(t) = \frac{\lambda(\lambda t)^{K-1}}{\Gamma(K)} \exp(-\lambda t), \quad t \geq 0, K \rangle 0, \lambda \rangle 0 \tag{2.49}$$

where

t is time.

K is the shape parameter.

$\Gamma(K)$ is the gamma function.

$\lambda = \frac{1}{\theta}, \theta$ is the scale parameter.

Using Equation (2.49) in Equation (2.35) yields the following equation for the cumulative distribution function:

$$F(t) = 1 - \frac{\Gamma(K, \lambda t)}{\Gamma(K)} \tag{2.50}$$

where

$\Gamma(K, \lambda t)$ is the incomplete gamma function.

Substituting Equation (2.49) into Equation (2.37) we get the following equation for the distribution mean:

$$E(t) = \frac{K}{\lambda} \tag{2.51}$$

Three special case distributions of the gamma distribution are the exponential distribution, the chi-square distribution, and the special case Erlangian distribution [16].

2.6.5 Exponential Distribution

This is probably the most widely used statistical distribution in reliability studies because it is easy to handle in performing reliability analysis and many engineering items exhibit constant failure rates during their useful life [17]. Its probability density function is defined by

$$f(t) = \lambda e^{-\lambda t}, \qquad \lambda > 0,\ t \geq 0 \tag{2.52}$$

where

t is time.

λ is the distribution parameter. In reliability work, it is called constant failure rate.

Substituting Equation (2.52) into Equation (2.35) we get the following equation for the cumulative distribution function:

$$F(t) = 1 - e^{-\lambda t} \tag{2.53}$$

Using Equation (2.53) in Equation (2.37) yields the following equation for the distribution mean:

$$E(t) = \frac{1}{\lambda} \tag{2.54}$$

2.6.6 Rayleigh Distribution

The Rayleigh distribution is named after John Rayleigh (1842–1919) and is often used in the theory of sound and in reliability studies. The distribution probability density function is defined by

$$f(t) = \left(\frac{2}{\theta^2}\right) t \, e^{-\left(\frac{t}{\theta}\right)^2}, \qquad \theta \rangle 0, t \geq 0 \tag{2.55}$$

where
t is time.
θ is the distribution parameter.

By substituting Equation (2.55) into Equation (2.35) we get the following equation for the cumulative distribution function:

$$F(t) = 1 - e^{-\left(\frac{t}{\theta}\right)^2} \tag{2.56}$$

Inserting Equation (2.55) into Equation (2.37) yields the following equation for the distribution mean:

$$E(t) = \theta \Gamma\left(\frac{3}{2}\right) \tag{2.57}$$

where
$\Gamma(.)$ is the gamma function, which is expressed by

$$\Gamma(y) = \int_0^\infty t^{y-1} \, e^{-t} \, dt, \qquad y \rangle 0 \tag{2.58}$$

2.6.7 Weibull Distribution

The Weibull distribution is named after W. Weibull, a Swedish mechanical engineering professor who developed it in the early 1950s [17]. It is often used in reliability studies, and its probability density function is defined by

$$f(t) = \frac{\beta t^{\beta-1}}{\theta^\beta} \, e^{-\left(\frac{t}{\theta}\right)^\beta}, \qquad \theta \rangle 0, \beta \rangle 0, t \geq 0 \tag{2.59}$$

where
t is time.
θ and β are the distribution scale and shape parameters, respectively.

Using Equation (2.59) in Equation (2.35) yields the following equation for the cumulative distribution function:

$$F(t) = 1 - e^{-\left(\frac{t}{\theta}\right)^{\beta}} \tag{2.60}$$

By inserting Equation (2.59) into Equation (2.37) we get the following equation for the distribution mean:

$$E(t) = \theta\Gamma\left(1 + \frac{1}{\beta}\right) \tag{2.61}$$

It is to be noted that exponential and Rayleigh distributions are the special cases of this distribution for $\beta = 1$ and $\beta = 2$, respectively.

2.7 Problems

1. What is mean deviation?
2. Obtain the Laplace transform of the following function:

$$f(t) = \lambda e^{-\lambda t} \tag{2.62}$$

where
λ is a constant.
t is time.

3. Find roots of the following equation by using the quadratic formula:

$$x^2 + 15x + 50 = 0 \tag{2.63}$$

Approximate the real roots of the following equation by using the Newton method:

$$x^2 - 37 = 0 \tag{2.64}$$

4. Write down the five most important probability properties.
5. Prove that the total area under a continuous random variable probability density function curve is equal to unity.
6. Define the probability density function of a continuous random variable.
7. What are the special case distributions of the Weibull distribution?
8. Prove that the mean or the expected value of the gamma distribution is given by Equation (2.51).
9. Prove Equation (2.60).

References

1 Eves, H., An Introduction to the History of Mathematics, Holt, Rinehart and Winston, New York, 1976.
2 Owen, D.B., Editor, On the History of Statistics and Probability, Marcel Dekker, New York, 1976.
3 Spiegel, M.R., Statistics, McGraw Hill Book Company, New York, 1961.
4 Oberhettinger, F., Badic, L., Tables of Laplace Transforms, Springer-Verlag, New York, 1973.
5 Spiegel, M.R., Laplace Transforms, McGraw Hill Book Company, New York, 1965.
6 Swokowski, E.W., Calculus with Analytic Geometry, Prindle, Weber, and Schmidt, Boston, Massachusetts, 1979.
7 Scheid, F., Numerical Analysis, McGraw Hill Book Company, New York, 1968.
8 Lipschutz, S., Set Theory, McGraw Hill Book Company, New York, 1964.
9 Fault Tree Handbook, Report No. NUREG-0492, U.S. Nuclear Regulatory Commission, Washington, D.C., 1981.
10 Lipschutz, S., Probability, McGraw Hill Book Company, New York, 1965.
11 Mann, N.R., Schafer, R.E., Singpurwalla, N.D., Methods for Statistical Analysis of Reliability and Life Data, John Wiley and Sons, New York, 1974.
12 Ang, A.H.S., Tang, W.H., Probability Concepts in Engineering, John Wiley and Sons, New York, 2006.
13 Dhillon, B.S., Design Reliability: Fundamentals and Applications, CRC Press, Boca Raton, Florida, 1999.
14 Patel, J.K., Kapadia, C.H., Owen, D.W., Handbook of Statistical Distributions, Marcel Dekker, New York, 1976.
15 Gupta, S., Groll, P., Gamma Distribution in Acceptance Sampling Based on Life Tests, Journal of American Statistical Association, December 1961, pp. 942–970.
16 Dhillon, B.S., Mechanical Reliability: Theory, Models, and Applications, American Institute of Aeronautics and Astronautics, Washington, D.C., 1988.
17 Weibull, W., A Statistical Distribution Function of Wide Applicability, J. Appl. Mech., Vol. 18, 1951, pp. 293–297.

3

Introduction to Reliability and Quality

3.1 Introduction

Today the reliability of engineering systems has become an important factor during their planning, design, and operation. The factors that are responsible for this include high acquisition cost, increasing number of reliability-related lawsuits, complex and sophisticated systems, competition, public pressures, and the past well-publicized system failures. Needless to say, over the past 60 years, many new advances have been made in the field of reliability that help to produce reliable systems.

The importance of quality in business and industry has increased to a level greater than ever before. Factors such as growing demand from customers for better quality, the global economy, and the complexity and sophistication of products have played an instrumental role in increasing this importance. As per [1,2], the cost of quality control accounts for roughly 7–10% of the total sales revenue of manufacturers. Today, the industrial sector is faced with many quality-related challenges. Some of these are the rising cost of quality, the Internet economy, an alarming rate of increase in customer quality-related requirements, and the need for improvements in methods and practices associated with quality-related activities. This chapter presents various introductory aspects of both reliability and quality.

3.2 Bathtub Hazard Rate Concept and Reliability Basic Formulas

The bathtub hazard rate concept is widely used to represent failure behavior of many engineering items. The term "bathtub" stems from the fact that the shape of the hazard rate curve resembles a bathtub (Figure 3.1.) As shown in the figure, the curve is divided into three distinct regions: burn-in, useful life, and wear-out. During the burn-in region the item hazard rate decreases with time. Some of the

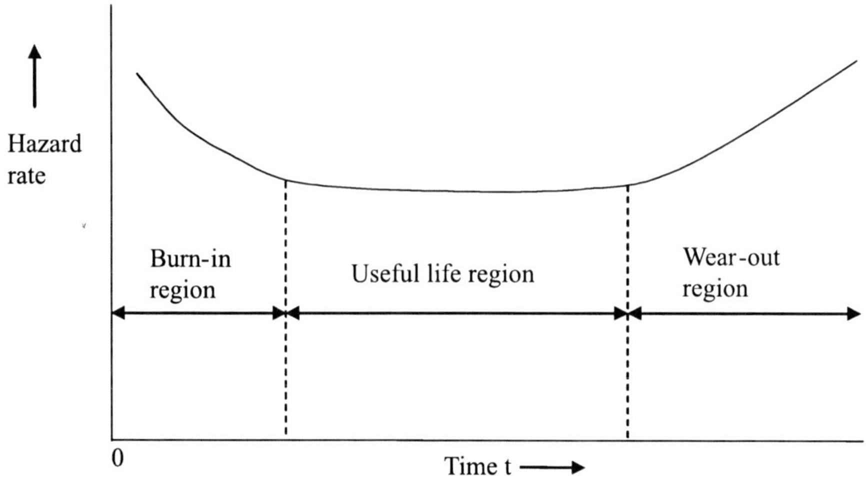

Figure 3.1. Bathtub hazard rate curve

reasons for the occurrence of failures during this region are poor quality control, poor manufacturing methods and procedures, poor debugging, poor workmanship and substandard materials, inadequate processes, and human error.

During the useful life region the item hazard rate remains constant with respect to time. Some of the main reasons for the occurrence of failures during this region are undetectable defects, higher random stress than expected, abuse, low safety factors, and human error.

During the wear-out region the item hazard rate increases with time. Some of the principal reasons for the occurrence of failures during this region are inadequate maintenance, wear due to aging, wear due to friction, short designed-in life of items, wrong overhaul practices, and corrosion and creep.

There are many basic formulas used in reliability work. Four widely used such formulas are as follows [3]:

$$f(t) = -\frac{\mathrm{d}R(t)}{\mathrm{d}t} \tag{3.1}$$

$$\lambda(t) = -\frac{1}{R(t)} \cdot \frac{\mathrm{d}R(t)}{\mathrm{d}t} \tag{3.2}$$

$$R(t) = \mathrm{e}^{-\int_0^t \lambda(t)\mathrm{d}t} \tag{3.3}$$

and

$$MTTF = \int_0^{\infty} R(t)\,\mathrm{d}t \tag{3.4}$$

where

t	is time.
$f(t)$	is the item failure (or probability) density function.
$R(t)$	is the item reliability at time t.
$\lambda(t)$	is the item hazard rate or time dependent failure rate.
$MTTF$	is the item mean time to failure.

Example 3.1

Assume that the failure rate, λ, of an engineering system is 0.0004 failures per hour. Calculate the following:

- System reliability during a 100-hour mission.
- System mean time to failure.

By substituting the specified data values into Equation (3.3), we get

$$R(100) = e^{-\int_0^{100} (0.0004)\,dt}$$

$$= e^{-(0.0004)(100)}$$

$$= 0.9608$$

Similarly, using the given data in Equation (3.4) yields

$$MTTF = \int_0^{\infty} e^{-(0.0004)t}\,dt$$

$$\int_0^{\infty} e^{ax}\,dx = \frac{-1}{a}$$

$$= \frac{1}{0.0004}$$

$$= 2500 \text{ } hours$$

Thus, the system reliability and mean time to failure are 0.9608 and 2500 hours, respectively.

Example 3.2

Assume that the hazard rate of a system is defined by

$$\lambda(t) = \frac{\alpha t^{\alpha-1}}{\theta^{\alpha}} \qquad (3.5)$$

where

α	is the shape parameter.
θ	is the scale parameter.
t	is time.

Obtain an expression for the system reliability.

By inserting Equation (3.5) into Equation (3.3), we get

$$\begin{aligned} R(t) &= e^{-\int_0^t \frac{\alpha t^{\alpha-1}}{\theta^\alpha} dt} \\ &= e^{-\left(\frac{t}{\theta}\right)^\alpha} \end{aligned} \tag{3.6}$$

Thus, Equation (3.6) is the expression for the system reliability.

3.3 Reliability Evaluation of Standard Configurations

As engineering systems can form various types of configurations in performing reliability analysis, this section presents reliability analysis of some standard networks or configurations.

3.3.1 Series Configuration

In this case, all units must work normally for the system success. A block diagram representing an m-unit series system is shown in Figure 3.2. Each block in the diagram represents a unit.

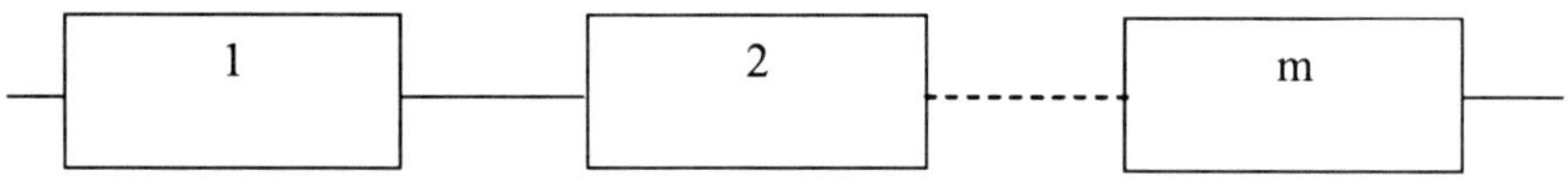

Figure 3.2. Block diagram of an m-unit series system

For independently failing units, the reliability of the series system shown in Figure 3.2 is

$$R_s = \prod_{i=1}^{m} R_i \tag{3.7}$$

where

R_s is the series system reliability.

m is the total number of units in series.

R_i is the unit i reliability; for $i = 1, 2, \ldots., m$.

For constant failure rate of unit i (*i. e.*, $\lambda_i(t) = \lambda_i$) from Equation (3.3), we get

$$\begin{aligned} R_i(t) &= e^{-\int_0^t \lambda_i dt} \\ &= e^{-\lambda_i t} \end{aligned} \tag{3.8}$$

where
$R_i(t)$ is the reliability of unit i at time t.
$\lambda_i(t)$ is the unit i hazard rate.
λ_i is the unit i constant failure rate.

By substituting Equation (3.8) into Equation (3.7), we get

$$
\begin{aligned}
R_s(t) &= \prod_{i=1}^{m} e^{-\lambda_i t} \\
&= e^{-\sum_{i=1}^{m} \lambda_i t}
\end{aligned}
\tag{3.9}
$$

where
$R_s(t)$ is the series system reliability at time t.

Using Equation (3.9) in Equation (3.4) yields

$$
\begin{aligned}
MTTF_s &= \int_0^{\infty} e^{-\sum_{i=1}^{m} \lambda_i t}\, dt \\
&= \frac{1}{\sum_{i=1}^{m} \lambda_i}
\end{aligned}
\tag{3.10}
$$

where
$MTTF_s$ is the series system mean time to failure.

Example 3.3
Assume that an aircraft has four independent and identical engines and all must work normally for the aircraft to fly successfully. Calculate the reliability of the aircraft flying successfully, if each engine's reliability is 0.99.

By substituting the given data values into Equation (3.7), we get

$$
\begin{aligned}
R_s &= (0.99)^4 \\
&= 0.9606
\end{aligned}
$$

Thus, the reliability of the aircraft flying successfully is 0.9606.

3.3.2 Parallel Configuration

In this case, the system is composed of m active units, and at least one such unit must operate normally for the system success. The system block diagram is shown in Figure 3.3. Each block in the diagram represents a unit.

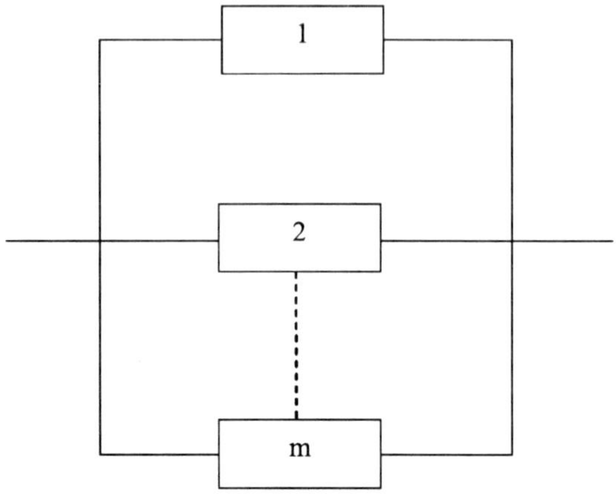

Figure 3.3. A parallel system with m units

For independently failing units, the parallel system reliability is given by

$$R_{ps} = 1 - \prod_{i=1}^{m} (1 - R_i) \tag{3.11}$$

where
R_{ps} is the parallel system reliability.
m is the total number of units in parallel.
R_i is the unit i reliability; for $i = 1, 2, \ldots, m$.

For constant failure rate, λ_i, of unit i by substituting Equation (3.8) into Equation (3.11), we get

$$R_{ps}(t) = 1 - \prod_{i=1}^{m} \left(1 - e^{-\lambda_i t}\right) \tag{3.12}$$

where
$R_{ps}(t)$ is the parallel system reliability at time t.

For identical units, inserting Equation (3.12) into Equation (3.4) yields.

$$\begin{aligned} MTTF_{ps} &= \int_0^\infty \left[1 - \left(1 - e^{-\lambda t}\right)^m\right] dt \\ &= \frac{1}{\lambda} \sum_{i=1}^{m} \frac{1}{i} \end{aligned} \tag{3.13}$$

where
λ is the unit constant failure rate.
$MTTF_{ps}$ is the parallel system mean time to failure.

Example 3.4
A system is composed of two independent and identical active units and at least one unit must operate normally for the system success. Each unit's constant failure rate is 0.0008 failures per hour. Calculate the system mean time to failure and reliability for a 150-hour mission.

Substituting the given data values into Equation (3.13) yields

$$MTTF_{ps} = \frac{1}{(0.0008)}\left(1+\frac{1}{2}\right) = 1875 \text{ hours}$$

Using the specified data values in Equation (3.12) yields

$$R_{ps}(150) = \left[1-\left\{1-e^{-(0.0008)(150)}\right\}^2\right]$$
$$= 0.9872$$

Thus, the system mean time to failure and reliability are 1875 hours and 0.9872, respectively.

3.3.3 *K*-out-of-*m* Configuration

In this case, the system is composed of m active units and at least K such units must work normally for the system success. The series and parallel configurations are special cases of this configuration for $K = m$ and $K = 1$, respectively.

For independent and identical units, the K-out-of-m configuration reliability is given by

$$R_{K/m} = \sum_{i=K}^{m} \binom{m}{i} R^i (1-R)^{m-i} \tag{3.14}$$

where

$$\binom{m}{i} = \frac{m!}{(m-i)!i!}$$

$R_{K/m}$ is the K-out-of-m configuration reliability.

R is the unit reliability.

For constant failure rates of units, using Equation (3.8) and (3.14), we get

$$R_{K/m}(t) = \sum_{i=K}^{m} \binom{m}{i} e^{-i\lambda t} \left(1-e^{-\lambda t}\right)^{m-i} \tag{3.15}$$

where

$R_{K/m}(t)$ is the K-out-of-m configuration reliability at time t.

λ is the unit constant failure rate.

Substituting Equation (3.15) into Equation (3.4) yields

$$\begin{aligned} MTTF_{K/m} &= \int_0^\infty \left[\sum_{i=K}^{m} \binom{m}{i} e^{-i\lambda t} \left(1 - e^{-\lambda t}\right)^{m-i} \right] dt \\ &= \frac{1}{\lambda} \sum_{i=K}^{m} \frac{1}{i} \end{aligned} \tag{3.16}$$

Example 3.5

Assume that a system is composed of three active, independent, and identical units, and at least two units must work normally for the system success. Calculate the system mean time to failure, if each unit's failure rate is 0.0004 failures per hour.

By substituting the specified data values into Equation (3.16), we get

$$\begin{aligned} MTTF_{K/m} &= \frac{1}{(0.0004)} \sum_{i=2}^{3} \frac{1}{i} \\ &= 2083.3 \text{ hours} \end{aligned}$$

Thus, the system mean time to failure is 2083.3 hours.

3.3.4 Standby System

In the case of the standby system, only one unit operates and m units are kept in their standby mode. As soon as the operating unit fails, the switching mechanism detects the failure and turns on one of the standbys. The system contains a total of $(m+1)$ units and it fails when all the m standby units fail. For perfect switching mechanism and standby units, independent and identical units, and units' constant failure rates, the standby system reliability is given by [4]

$$R_{std}(t) = \sum_{i=0}^{m} \frac{(\lambda t)^i e^{-\lambda t}}{i!} \tag{3.17}$$

where

$R_{std}(t)$ is the standby system reliability at time t.

m is the total number of standby units.

λ is the unit constant failure rate.

Using Equation (3.17) in Equation (3.4) yields

$$\begin{aligned} MTTF_{std} &= \int_0^\infty \left[\sum_{i=0}^{m} \frac{(\lambda t)^i e^{-\lambda t}}{i!} \right] dt \\ &= \frac{m+1}{\lambda} \end{aligned} \tag{3.18}$$

where

$MTTF_{\text{std}}$ is the standby system mean time to failure.

Example 3.6

A system has two independent and identical units. One of these units is operating, and the other is on standby. Calculate the system mean time to failure and reliability for a 200-hour mission by using Equations (3.17) and (3.18), if the unit failure rate is 0.0001 failures per hour.

By substituting the given data values into Equation (3.17), we get

$$R_{\text{std}}(200) = \sum_{i=0}^{1} \frac{\left[(0.0001)(200)\right]^i e^{-(0.0001)(200)}}{i!}$$
$$= 0.9998$$

Similarly, substituting the given data values into Equation (3.18) yields

$$MTTF_{\text{std}} = \frac{(1+1)}{(0.0001)}$$
$$= 20,000 \text{ hours}$$

Thus, the system reliability and mean time to failure are 0.9998 and 20,000 hours, respectively.

3.3.5 Bridge Configuration

In some engineering systems, particularly communications networks, units may form a bridge configuration (Figure 3.4). Each block in the figure represents a unit, and numerals in blocks denote unit number. For independent units, the reliability of the Figure 3.4 bridge configuration is [5]

$$\begin{aligned} R_b = {} & 2R_1 R_2 R_3 R_4 R_5 + R_2 R_3 R_4 + R_1 R_3 R_5 + R_2 R_5 + R_1 R_4 \\ & - R_2 R_3 R_4 R_5 - R_1 R_2 R_3 R_4 - R_1 R_2 R_3 R_5 \\ & - R_1 R_3 R_4 R_5 - R_1 R_2 R_4 R_5 \end{aligned} \tag{3.19}$$

where

R_i is the unit i reliability; for $i = 1, 2, 3, 4, 5$.

R_b is the bridge configuration or system reliability.

For identical units, Equation (3.19) reduces to

$$R_b = 2R^5 - 5R^4 + 2R^3 + 2R^2 \tag{3.20}$$

where

R is the unit reliability.

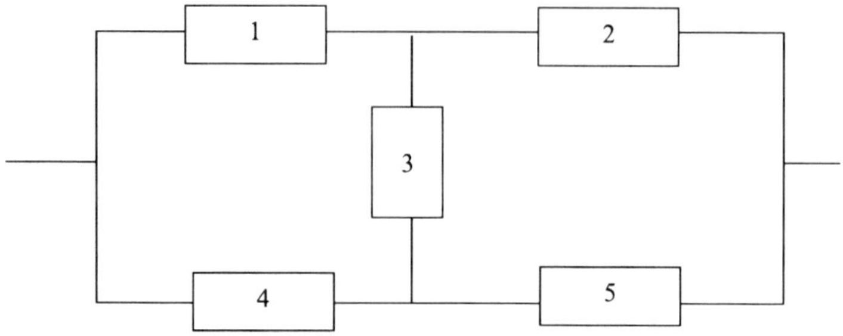

Figure 3.4. A five-unit bridge network

For constant failure rates of units, using Equations (3.8) and (3.20), we get

$$R_b(t) = 2e^{-5\lambda t} - 5e^{-4\lambda t} + 2e^{-3\lambda t} + 2e^{-2\lambda t} \tag{3.21}$$

Inserting Equation (3.21) into Equation (3.4) yields

$$\begin{aligned} MTTF_b &= \int_0^\infty \left[2e^{-5\lambda t} - 5e^{-4\lambda t} + 2e^{-3\lambda t} + 2e^{-2\lambda t}\right] dt \\ &= \frac{49}{60\lambda} \end{aligned} \tag{3.22}$$

where

$MTTF_b$ is the bridge system mean time to failure.

λ is the unit constant failure rate.

Example 3.7

Assume that five independent and identical units form a bridge configuration and each unit's reliability is 0.9. Calculate the bridge configuration reliability.

By substituting the given data value into Equation (3.20), we get

$$\begin{aligned} R_b &= 2(0.9)^5 - 5(0.9)^4 + 2(0.9)^3 + 2(0.9)^2 \\ &= 0.9785 \end{aligned}$$

Thus, the bridge configuration reliability is 0.9785.

3.4 Reliability Analysis Methods

There are many methods that can be used to perform reliability analysis of engineering systems [4, 5]. This section presents three of these commonly used methods.

3.4.1 Failure Modes and Effect Analysis (FMEA)

Failure modes and effect analysis (FMAE) is a widely used method in the industrial sector to perform reliability analysis of engineering systems. It may simply be described as an approach used to conduct analysis of each potential failure mode in the system under consideration to examine the effects of such failure modes on that system [6]. Furthermore, the approach demands listing of all potential failure modes of all parts on paper and their effects on the listed subsystems and the system under consideration. When this approach (*i. e.*, FMEA) is extended to classify each potential failure effect according to its severity, then it is called failure mode, effects, and criticality analysis (FMECA) .

The history of FMEA goes back to the early 1950s, when the U.S. Navy's Bureau of Aeronautics developed a requirement called "Failure Analysis" [7]. In the 1970s, the U.S. Department of Defense developed a military standards entitled "Procedures for Performing a Failure Mode, Effects, and Criticality Analysis" [8]. FMEA is described in detail in [5] and a comprehensive list of publications on FMEA/FMECA is available in [9].

The six main steps followed in performing FMEA are shown in Figure 3.5 [10]. Some of the main characteristics of the FMEA method are as follows:

- It is a routine upward approach that starts from the detailed level.
- By determining all possible failure effects of each part, the entire system is screened completely.
- It identifies weak spots in a system design and highlights areas where further or detailed analyses are required.
- It improves communication among individuals involved in design.

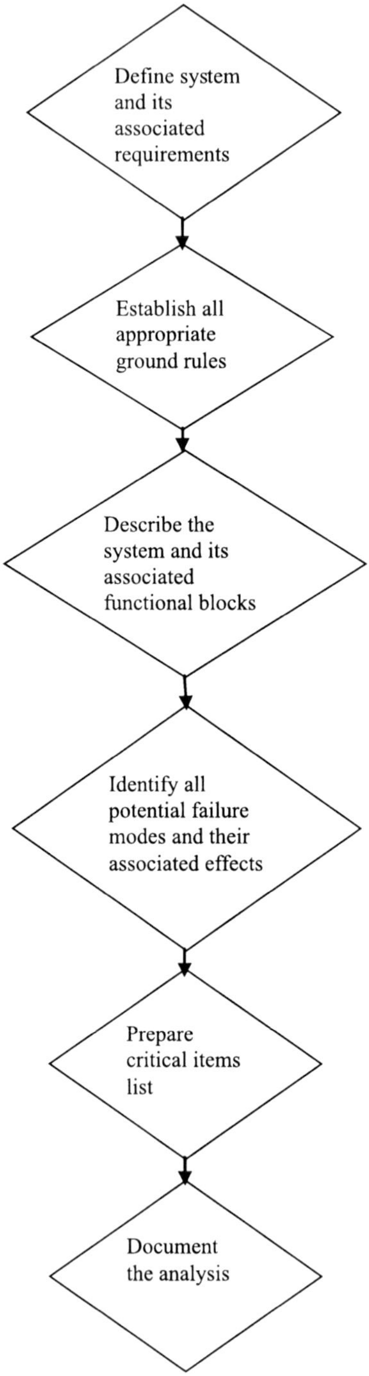

Figure 3.5. Main steps for conducting FMEA

3.4.2 Markov Method

The Markov method is a powerful reliability analysis tool that is named after Russian mathematician Andrei Andreyevich Markov (1856–1922). It can handle both repairable and non-repairable systems. In analyzing large and complex systems, a problem may occur in solving a set of differential equations generated by this method. Nonetheless, the Markov method is based on the following assumptions [4]:

- The transitional probability from one state to the next state in the finite time interval Δt is given by $\lambda \Delta t$, where λ is transition rate (*e. g.*, failure or repair rate) associated with Markov states.
- The probability of more than one transition occurrence in finite time interval Δt from one state to the next is negligible (*e. g.*, $(\lambda \Delta t)\,(\lambda \Delta t) \rightarrow 0$).
- All occurrences are independent of each other.

The application of this method is demonstrated through the following example.

Example 3.8
Assume that a system can either be in an operating or a failed state and its failure rate, λ_S, is constant. The system state space diagram is shown in Figure 3.6. The numerals in boxes denote the system state. Obtain expressions for system state probabilities (*i. e.*, system operating or failed) by using the Markov method.

Using the Markov method, we write down the following two equations for the diagram in Figure 3.6:

$$P_0\left(t+\Delta t\right) = P_0(t)\left(1-\lambda_s \Delta t\right) \tag{3.23}$$

$$P_1\left(t+\Delta t\right) = P_1(t) + P_0(t)\lambda_s \Delta t \tag{3.24}$$

where

$P_i\,(t+\Delta t)$ is the probability that the system is in state i at time $(t+\Delta t)$; for $i=0$ (operating normally), $i=1$ (failed).

$P_i\,(t)$ is the probability that the system is in state i at time t; for $i=0$ (operating normally), $i=1$ (failed).

λs is the system constant failure rate.

$\lambda_s\,\Delta t$ is the probability of system failure in finite time interval Δt.

$(1-\lambda_s\,\Delta t)$ is the probability of no failure in time interval Δt when the system is in state 0.

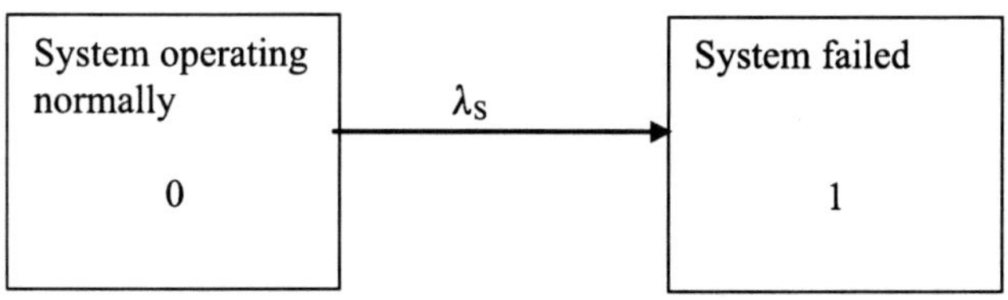

Figure 3.6. System state space diagram

In the limiting case, Equations (3.23) and (3.24) become

$$\lim_{\Delta t \to 0} \frac{P_0(t+\Delta t) - P_0(t)}{\Delta t} = \frac{dP_0(t)}{dt} = -\lambda_s P_0(t) \tag{3.25}$$

and

$$\lim_{\Delta t \to 0} \frac{P_1(t+\Delta t) - P_1(t)}{\Delta t} = \frac{dP_1(t)}{dt} = -\lambda_s P_0(t) \tag{3.26}$$

At time $t=0$, $P_0(0)=1$, and $P_1(0)=0$.

Solving Equations (3.25) and (3.26) by using Laplace transforms, we get

$$P_0(s) = \frac{1}{s+\lambda_s} \tag{3.27}$$

and

$$P_1(s) = \frac{1}{s(s+\lambda_s)} \tag{3.28}$$

where

s is the Laplace transform variable.

$P_i(s)$ is the Laplace transform of the probability that the system is in state i; for $i=0$ (operating normally), $i=1$ (failed).

Taking the inverse Laplace transforms of Equations (3.27) and (3.28), we get

$$P_0(t) = e^{-\lambda_s t} \tag{3.29}$$

$$P_1(t) = 1 - e^{-\lambda_s t} \tag{3.30}$$

Example 3.9

Assume that the constant failure rate of a system is 0.002 failures per hour. Calculate the system probability of failure during a 150-hour mission.

By substituting the given data values into Equation (3.30), we get

$$P_1(150) = 1 - e^{-(0.002)(150)}$$
$$= 0.2592$$

It means, the system probability of failure is 0.2592.

3.4.3 Fault Tree Analysis (FTA)

Fault tree analysis (FTA) is one of the most widely used methods in the industrial sector to evaluate reliability of engineering systems. The method was developed in the early 1960s at Bell Telephone Laboratories to evaluate the reliability and safety of the minuteman Launch Control System [11]. The method is described in detail in [11].

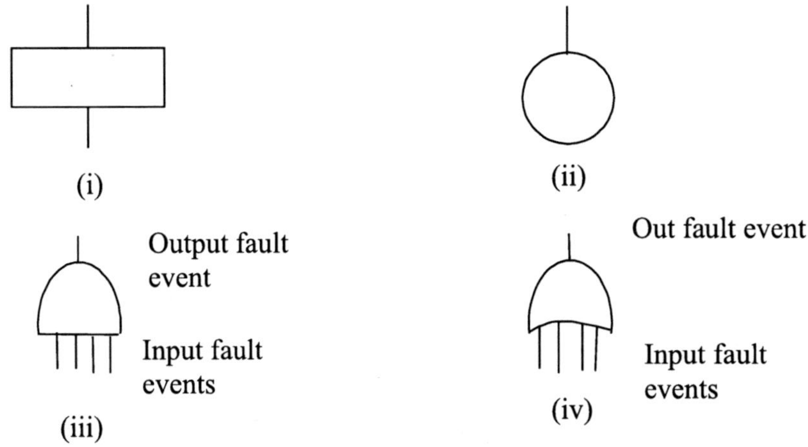

Figure 3.7. Commonly used fault tree symbols: (**i**) rectangle, (**ii**) circle; (**iii**) AND gate; (**iv**) OR gate.

Although many symbols are used in performing FTA, the four commonly used symbols are shown in Figure 3.7. Each of these symbols is described below.

- **AND gate.** This denotes that an output fault event occurs only if all of the input fault events occur.
- **OR gate.** This denotes that an output fault event occurs if one or more of the input fault events occur.
- **Rectangle.** This denotes a fault event that results from the logical combination of fault events through the input of a logic gate.
- **Circle.** This represents a basic fault event or the failure of an elementary component. The event's probability of occurrence, failure, and repair rates are normally obtained from field failure data.

FTA begins by identifying an undesirable event, called a top event, associated with a system. Fault events which could cause the top event are generated and connected by logic gates such as OR and AND. The fault tree construction proceeds by generation of fault events in a successive manner until the events need not be developed any further.

Example 3.10

Assume that a windowless room contains one switch and four light bulbs and the switch can only fail to close. Develop a fault tree for the top event "Dark room" (*i. e.*, no light in the room), if the interruption of electrical power coming into the room can only be caused either by fuse failure or power failure.

By using the Figure 3.7 symbols, a fault tree for the example shown in Figure 3.8 is developed. Each fault event in the figure is labeled as X_0, X_1, X_2, X_3, X_4, X_5, X_6, X_7, X_8, and X_9.

Probability Evaluation of Fault Trees

For independent fault events, the probability of occurrence of top events of fault trees can easily be evaluated by applying the basic rules of probability to the output fault events of logic gates. For example, in the case of Figure 3.8 fault tree, we have [5]

$$P(X_2) = P(X_6)P(X_7)P(X_8)P(X_9) \tag{3.31}$$

$$P(X_1) = P(X_4) + P(X_5) - P(X_4)P(X_5) \tag{3.32}$$

$$P(X_0) = 1 - [1 - P(X_1)][1 - P(X_2)][1 - P(X_3)] \tag{3.33}$$

where
$P(X_i)$ is the probability of occurrence of fault event X_i; for $i = 1, 2, 3, \ldots, 9$.

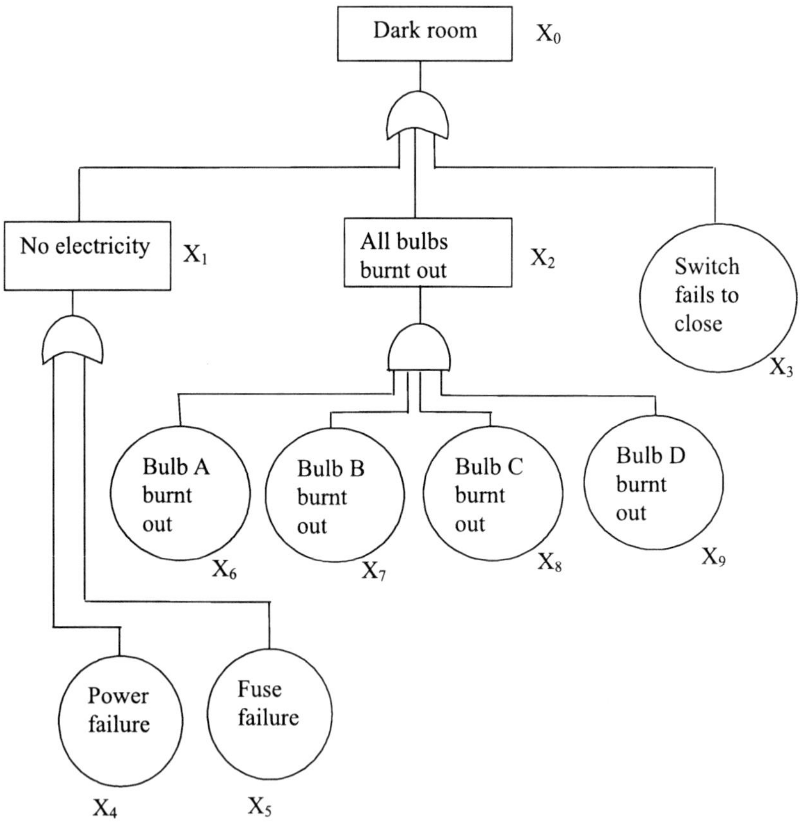

Figure 3.8. A fault tree for Example 3.10

Example 3.11

In Figure 3.8, assume that the probability of occurrence of fault events X_3, X_4, X_5, X_6, X_7, X_8, and X_9 are 0.01, 0.02, 0.03, 0.04, 0.05, 0.06, and 0.07, respectively. Calculate the probability of occurrence of the top event "Dark room" by using Equations (3.31)–(3.33).

Thus, by substituting the given data values into Equations (3.31)–(3.33), we get

$$P(X_2) = (0.04)(0.05)(0.06)(0.07) = 0.0000084$$

$$P(X_1) = 0.02 + 0.03 - (0.02)(0.03) = 0.0494$$

and

$$P(X_0) = 1 - (1 - 0.0494)(1 - 0.0000084)(1 - 0.01) \\ = 0.9411$$

Thus, the probability of occurrence of the top event "Dark room" is 0.9411.

3.5 Quality Goals, Quality Assurance System Elements, and Total Quality Management

Normally, organizations set various types of quality goals. These goals may be divided under the following two distinct categories [12]:

- **Goals for breakthrough.** These are concerned with improving the existing quality of products or services. Three important reasons for establishing such goals are shown in Figure 3.9.
- **Goals for control.** These goals are concerned with maintaining the quality of products or services to the current level for a specified period of time. Some of the important reasons for establishing such goals are as follows:
  - Acceptable competitiveness at current levels of quality
  - Improvements are uneconomical
  - Insignificant number of customers or other complaints about the quality of products or services

The main goal of a quality assurance system is to maintain the specified level of quality, and its important elements are as follows [13]:

- Evaluate, plan, and control product quality.
- Consider the quality and reliability needs during the product design and development.
- Keep track of suppliers' quality assurance programs.
- Develop personnel.
- Determine and control product quality in use environment.
- Conduct special quality studies.

- Provide quality-related information to management.
- Assure accuracy of quality measuring equipment.
- Manage the total quality assurance system.

The term total quality management (TQM) was coined by Nancy Warren, a behavioral scientist [14]. Some of the important elements of TQM are management commitment and leadership, team work, customer service, quality cost, training, statistical approaches, and supplier participation [15]. For the success of the TQM process, goals such as listed below must be satisfied in an effective manner [16].

- Clear understanding of all internal and external customer requirements by all company personnel
- Meeting of all control guidelines, per customer requirements, by all involved systems and processes
- Use of a system to continuously improve processes that better satisfy current and future needs of customers
- Establishment of appropriate incentives and rewards for employees when process control and customer satisfaction results are attained successfully

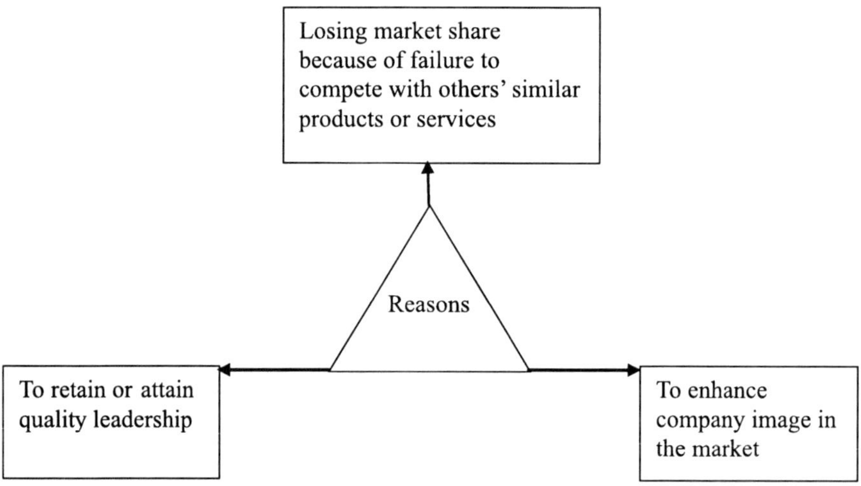

Figure 3.9. Reasons for establishing quality goals for breakthrough

3.6 Quality Analysis Methods

Over the years, many methods and techniques have been developed to conduct various types of quality-related analysis. This section presents some of these methods.

3.6.1 Quality Control Charts

A control chart may simply be described as a graphical method used for determining whether a process is in a "state of statistical control" or out of control [17]. The history of control charts may be traced back to a memorandum written by Walter Shewhart on May 16, 1924, in which he presented the idea of a control chart [18]. Nonetheless, the construction of control charts is based on statistical principles and distributions and a chart is basically composed of three elements: average or standard value of the characteristic under consideration, upper control limit (UCL), and lower control limit (LCL).

There are many types of quality control charts: the P-charts, the C-charts, the R-charts, the $\overline{X}$-charts, etc. [19, 20]. One of these is described below.

The P-charts

P-charts are also known as the control charts for attributes, in which the data population is grouped under two classifications (*e. g.*, pass or fail, good or bad). More specifically, components with defects and components without defects. Thus, attributes control charts use pass-fail information for charting and a p-chart basically is a single chart that tracks the proportion of nonconforming items in each sample taken from representative population.

Upper and lower control limits of p-charts are established by using the binomial distribution; thus are expressed by

$$UCL_p = m_b + 3\sigma_b \tag{3.34}$$

and

$$LCL_p = m_b - 3\sigma_b \tag{3.35}$$

where

m_b is the mean of the binomial distribution.
σ_b is the standard deviation of the binomial distribution.
UCL_p is the upper control limit of the p-chart.
LCL_p is the lower control limit of the p-chart.

The mean, m_b, is given by

$$m_b = \frac{K}{n\theta} \tag{3.36}$$

where
n is the sample size.
K is the total number of defectives/failures in classification.
θ is the number of samples.

Similarly, the standard deviation, σ_b, is given by

$$\sigma_b = \left[m_b \left(1 - m_b\right) / n \right]^{1/2} \tag{3.37}$$

Example 3.12
A total of eight samples were taken from the production line of a firm manufacturing mechanical components for use in a nuclear power plant. Each sample contained 60 components. The inspection process revealed that samples 1, 2, 3, 4, 5, 6, 7, and 8 contain 5,2, 12, 4, 8, 10, 15, and 6 defective components, respectively. Construct the p-chart for mechanical components.

Using the given data values in Equation (3.36) yields

$$m_b = \frac{(5+2+12+4+8+10+15+6)}{(60)(8)}$$
$$= 0.1292$$

By substituting the above calculated value and the other given data value into Equation (3.37), we get

$$\sigma_b = \left[0.1292(1-0.1292)/60\right]^{1/2}$$
$$= 0.0433$$

The fraction of defectives, p, in sample 1 is given by

$$p = \frac{5}{60} = 0.083$$

Similarly, the fractions of defective components in samples 2, 3, 4, 5, 6, 7, and 8 are 0.033, 0.2, 0.066, 0.133, 0.166, 0.25, and 0.1, respectively.

Substituting the above-calculated values for m_b and σ_b into Equations (3.34) and (3.35) yield

$$UCL_p = 0.1292 + 3(0.0433)$$
$$= 0.2591$$

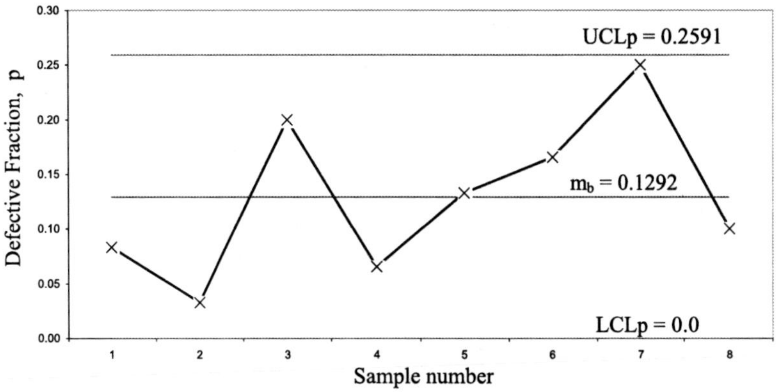

Figure 3.10. p-chart for mechanical components

and

$$\begin{aligned} LCL_p &= 0.1292 - 3(0.0433) \\ &= -0.0007 \\ &\cong 0.0 \end{aligned}$$

A p-chart for the above calculated values is shown in Figure 3.10. The crosses in the figure represent the fraction of defective components in each sample.

As all these crosses are within the upper and lower control limits, it means that there is no abnormality in the ongoing production process.

3.6.2 Cause-and-Effect Diagram

A Cause-and-effect diagram is basically a picture made up of lines and symbols designed to represent a meaningful relationship between an effect and its associated causes [21]. Other names used for this diagram or method are Ishikawa diagram (*i. e.*, after its originator, Kaoru Ishikawa) and "fishbone" diagram because of its resemblance to the skeleton of a fish. Nonetheless, the cause-and-effect diagram is a useful tool to determine the root causes of a given problem and generate relevant ideas.

From the quality perspective, the effect or problem could be a quality characteristic that needs improvement and the causes work methods, equipment, materials, people, environment, etc. Usually, the five steps shown in Figure 3.11 are followed to develop a cause-and-effect diagram.

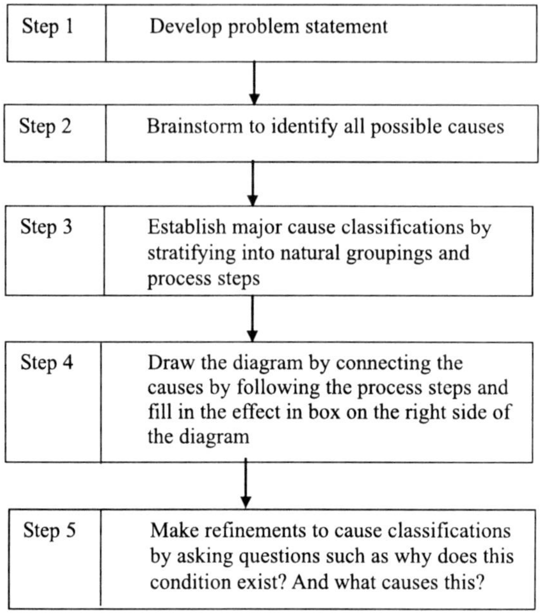

Figure 3.11. Steps for developing a cause-and-effect diagram

3.6.3 Quality Function Deployment (QFD)

The quality function deployment approach was developed in the 1960s in Japan and is used for optimizing the process of developing and manufacturing new products as per customer requirements [22–23]. Thus, QFD may simply be described as a formal process employed for translating customer needs into a set of technical requirements. The approach makes use of a set of matrices for relating customer requirements to counterpart characteristics that are expressed as process control requirements and technical specifications.

A QFD matrix is often referred to as the "House of Quality" because of its resemblance to the structure of a house. The main steps used to build the house of quality are as follows [22–24]:

- Highlight customer needs or requirements.
- Identify lay process/product characteristics that will meet customer requirements.
- Establish all necessary relationships between the customer needs and the counterpart characteristics.
- Analyse competing products.
- Establish all competing products' counterpart characteristics and develop appropriate goals.
- Identify counterpart characteristics to be utilized in the remaining process.

Additional information on this method is available in [24].

3.6.4 Pareto Diagram

The Pareto diagram is named after Wilfredo Pareto (1848–1923), an Italian economist and sociologist, and it may simply be described as a bar chart that ranks related problems/measures in decreasing occurrence frequency. In the quality area, the Pareto diagram or principle was first introduced by J.M. Juran, who believed that there are always a few types of defects in the hardware manufacture that loom large in the frequency of occurrence and severity [25, 26]. In other words, about 20% of the problems cause around 80% of the scrap. Usually, the steps followed to construct a Pareto diagram are shown in Figure 3.12 [21].

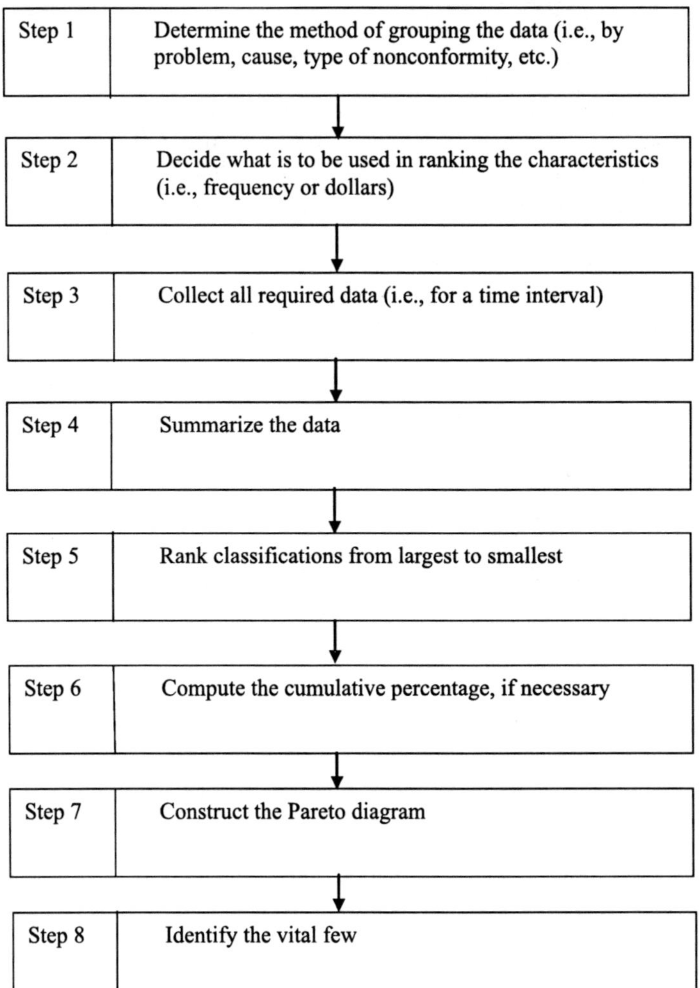

Figure 3.12. Steps for constructing the Pareto diagram

3.7 Quality Costs and Indices

Quality costs are a significant element of the sales income in many manufacturing organizations. They may be classified under five distinct categories as shown in Figure 3.13 [27–28]. These are administrative costs, appraisal and detection costs, prevention costs, internal failure costs, and external failure costs.

The administrative costs are concerned with administrative-related activities such as performing data analysis, reviewing contracts, preparing budgets, preparing proposals, and forecasting. The appraisal and detection costs are associated with appraisal and detection activities, and three main elements of these costs are cost of auditing, cost of inspection (*i. e.*, receiving, shipping, source, in-process, etc.), and cost of testing. The prevention costs are concerned with activities performed to prevent the production of defective products, parts, and materials. Some of these activities are reviewing designs, training personnel, evaluating suppliers, calibrating and certifying inspection and test devices and instruments, implementing and maintaining sampling plans, and coordinating plans and programs.

The internal failure costs occur prior to the delivery of the product to the buyer. They are associated with items such as in-house components and materials failures, redesign, scrap, failure analysis, and re-inspection and retest. The external failure costs occur after the delivery of the product to the buyers and are associated with items such as warranty charges, replacement of defective parts, liability, investigation of customer complaints, failure analysis, and repair.

Often, manufacturing organizations use various types of quality cost indices to monitor their performance. The values of such indices are plotted periodically and their trends are monitored. Three of these indices are as follows [29, 30].

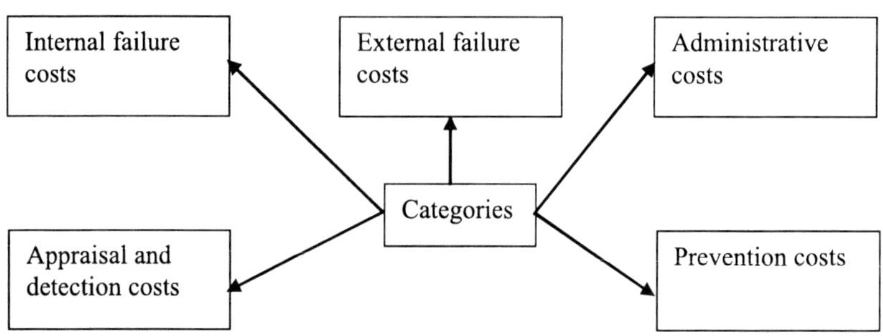

Figure 3.13. Quality cost categories

3.7.1 Index I

Index I is expressed by

$$\theta_1 = \left[\frac{QC_t(100)}{V} \right] + 100 \tag{3.38}$$

where
θ_1 is the quality cost index.
V is the value of output
QC_t is the total quality cost.

The common interpretations of three θ_1 values in the industrial sector are presented in Table 3.1 [30].

Table 3.1. Interpretations of three θ_1 values

θ_1 value	Interpretation
100	There is no defective output.
105	It can readily be achieved in a real-life environment.
110–130	Quality costs are ignored.

3.7.2 Index II

Index II is expressed by

$$\theta_2 = \frac{QC_t(100)}{LC_d} \tag{3.39}$$

where
θ_2 is the quality cost index expressed as a percentage.
LC_d is the direct labor cost.

Although this index does not provide management with much useful information for problem diagnosis and decision making, it is often used to eliminate the effects of inflation [30].

3.7.3 Index III

Index III is expressed by [30]

$$\theta_3 = \frac{QC_t(100)}{S_t} \tag{3.40}$$

where
θ_3 is the quality cost index expressed as a percentage.
S_t is the total sales.

3.8 Problems

1. Discuss the bathtub hazard rate curve.
2. Obtain a hazard rate expression for a series system by using Equations (3.2) and (3.9). Comment on the end result.
3. Prove the resulting Equation (3.13).
4. Assume that a system is composed of four active, independent, and identical units and at least two units must work normally for the system success. Calculate the system mean time to failure, if each unit's failure rate is 0.0005 failures per hour.
5. Compare standby system with the *k*-out-of-*m* configuration.
6. Compare FMEA with FTA.
7. Discuss at least eight important elements of a quality assurance system.
8. Describe the following two quality analysis methods:
   - Pareto diagram
   - Cause-and-effect diagram
9. What are the five categories of quality costs?
10. Who coined the term total quality management?

References

1 Feigenbaum, A.V., Total Quality Control, McGraw Hill Book Company, New York, 1983.
2 Dhillon, B.S., Quality Control, Reliability, and Engineering Design, Marcel Dekker, Inc., New York, 1985.
3 Grant Ireson, W., Coombs, C.F., Moss, R.Y., Editors, Handbook of Reliability Engineering and Management, McGraw Hill Book Company, New York, 1996.
4 Shooman, M.L., Probabilistic Reliability: An Engineering Approach, McGraw Hill Book Company, New York, 1968.
5 Dhillon, B.S., Design Reliability: Fundamentals and Applications, CRC Press, Boca Raton, Florida, 1999.
6 Omdahl, T.P., Editor, Reliability, Availability, and Maintainability (RAM) Dictionary, American Society for Quality Control (ASQC) Press, Milwaukee, Wisconsin, 1988.
7 MIL-F-18372 (Aer.), General Specification for Design, Installation, and Test of Air Flight Control Systems, Bureau of Naval Weapons, Department of the Navy, Washington, D.C.
8 MIL-STD-1629, Procedures for Performing a Failure Mode, Effects, and Criticality Analysis, Department of Defense, Washington, D.C.
9 Dhillon, B.S., Failure Mode and Effect Analysis: Bibliography, Microelectronics and Reliability, Vol. 32, 1992, pp. 719–731.
10 Jordan, W.E., Failure Modes, Effects, and Criticality Analyses, Proceedings of the Annual Reliability and Maintainability Symposium, 1972, pp. 30–37.

11 Dhillon, B.S., Singh, C., Engineering Reliability: New Technique and Applications, John Wiley and Sons, New York, 1981.
12 Juran, J.M., Gryna, F.M., Bingham, R.S., Quality Control Handbook, McGraw Hill Book Company, New York, 1979.
13 The Quality World of Allis-Chalmers, Quality Assurance, Vol. 9, 1970, pp. 13–17.
14 Walton, M., Deming Management at Work, Putnam, New York, 1990.
15 Burati, J.L., Matthews, M.F., Kalidindi, S.N., Quality Management Organization and Techniques, Journal of Construction Engineering and Management, Vol. 118, March 1992, pp. 112–128.
16 Dhillon, B.S., Reliability, Quality, and Safety for Engineers, CRC Press, Boca Raton, Florida, 2005.
17 Rosander, A.C., Applications of Quality Control in the Service Industries, Marcel Dekker, New York, 1985.
18 Juran, J.M., Early SQC: A Historical Supplement, Quality Progress, Vol. 30, No. 9, 1997, pp. 73–81.
19 Vance, L.C., A Bibliography of Statistical Quality Control Chart Techniques 1970–1980, Journal of Quality Technology, Vol. 15, No. 12, 1983, pp. 225–235.
20 Ryan, T.P., Statistical Methods for Quality Improvements, John Wiley and Sons, New York, 2000.
21 Besterfield, D. H., Quality Control, Prentice Hall, Upper Saddle River, New Jersey, 2001.
22 Akao, Y., Quality Function Deployment: Integrating Customer Requirements into Product Design, Productivity Press, Cambridge, Massachusetts, 1990.
23 Mizuno, S., Akao, Y., Editor, QFD: The Customer-Driven Approach to Quality Planning and Deployment, Asian Productivity Organization, Tokyo, 1994.
24 Bossert, J.L., Quality Function Deployment: A Practitioner's Approach, ASQC Quality Press, Milwaukee, Wisconsin, 1991.
25 Juran, J.M., Editor, Quality Control Handbook, McGraw Hill Book Company, New York, 1974, pp. 2.16–2.19.
26 Smith, G.M., Statistical Process Control and Quality Improvement, Prentice Hall, Upper Saddle River, New Jersey, 2001.
27 Hayes, G.E., Romig, H.G., Modern Quality Control, Collier-Macmillan, London, 1977.
28 Carter, C.L., The Control and Assurance of Quality, Reliability, and Safety, Published by C.L. Carter and Associates, Richardson, Texas, 1978.
29 Sullivan, E., Owens, D.A., Catching a Glimpse of Quality Costs Today, Quality Progress, Vol. 16, No. 12, 1983, pp. 21–24.
30 Evans, J.R., Lindsay, W.A., The Management and Control of Quality, West Publishing Company, New York, 1989.

4

Robot Reliability

4.1 Introduction

Robots are increasingly being used to perform various types of tasks including spot welding, materials handling, routing, and arc welding. A robot may simply be described as a mechanism guided by automatic controls. The word "robot" is derived from the Czechoslovakian language, in which it means "worker" [1]. In 1954, George Devol designed and applied for a patent for a programmable device that could be considered the first industrial robot. Nonetheless, in 1959, the Planet Corporation manufactured the first commercial robot [2]. Currently, the world-wide industrial robot population is estimated to be around one million [3].

As robots use mechanical, electrical, electronics, hydraulic, and pneumatic components, their reliability-related problems are quite challenging because of many different sources of failures. Although there is no clear-cut definitive point in the beginning of robot reliability and maintainability field, a publication by J.F. Engelberger, in 1974, could be regarded as its starting point [4]. In 1987, an article presented a comprehensive list of publications on robot reliability [5], and in 1991, a book entitled "Robot Reliability and Safety" covered the topic of robot reliability in a significant depth [6]. A comprehensive list of publications on robot reliability up to 2002 is available in Ref. [7], and some of the important recent publications on robot reliability and associated areas are listed at the end of this book. This chapter presents various important aspects of robot reliability.

4.2 Terms and Definitions

There are many robot reliability-related terms and definitions. Some of the important ones are as follows [1,6,8–12]:

- **Robot reliability.** This is the probability that a robot will perform its specified mission according to stated conditions for a given time period.
- **Robot availability.** This is the probability that a robot is available for service at the moment of need.
- **Graceful failure.** The performance of the manipulator degraded at a slow pace, in response to overloads, instead of failing catastrophically.
- **Erratic robot.** A robot moved appreciably off its specified path.
- **Robot mean time to failure.** This is the average time that a robot will operate before failure.
- **Robot mean time to repair.** This is the average time that a robot is expected to be out of operation after failure.
- **Fail-safe.** This is the failure of a robot/robot part without endangering people or damage to equipment or plant facilities.
- **Fault in teach pendant.** This is the part failure in the teach pendant of a robot.
- **Robot out of synchronization.** This is when the position of the robot's arm is not in line with the robot's memory of where it is supposed to be.
- **Error recovery.** This is the capability of intelligent robotic systems to reveal errors and, through programming, to initiate appropriate correction actions to overcome the impending problem and complete the specified process.
- **Robot repair.** This is to restore robots and their associated parts or systems to an operational condition after experiencing failure, damage, or wear.

4.3 Robot Failure Causes and Classifications

There are many causes of robot failures. Some of the most common ones are as follows [6]:

- Oil pressure valve problems
- Printed circuit board problems
- Human errors
- Encoder-related problems
- Servo valve problems
- Noise

As per Refs [13, 14], robot problems or troubles followed the following order:

- Control system problems
- Incompatibility of jigs and other tools
- Robot body-related problems
- Programming and operation errors
- Welding gun troubles and difficulties with other tooling parts

- Deterioration, precision deficiency
- Runaway
- Miscellaneous

There are basically four types of failures (Figure 4.1), that affect robot reliability and its safe operation [15, 16]. These are random component failures, systematic hardware faults, human errors, and software failures.

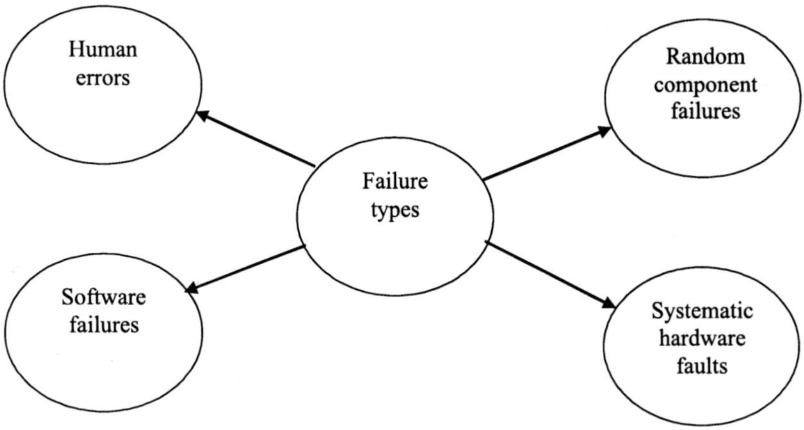

Figure 4.1. Types of failures that affect robot reliability

Failures that occur during the useful life of a robot are called random component failures because they occur unpredictably. Some of the reasons for the occurrence of such failures are undetectable defects, low safety factors, unexplainable causes, and unavoidable failures. Systematic hardware faults are those failures that occur because of the existence of unrevealed mechanisms in the robot design. Reasons such as peculiar wrist orientations and unusual joint-to-straight-line mode transition can lead the robot not to perform a specific task or to execute certain portions of a program.

Human errors are caused by people who design, manufacture, test, operate, and maintain robots. Various studies reveal that the human error is a significant element of total equipment failures [6, 17]. Some of the important reasons for the occurrence of human errors are poor equipment design, poorly trained operation and maintenance personnel, task complexity, inadequate lighting in the work areas, improper tools used by maintenance personnel, and poorly written maintenance and operating procedures [17]. Software failures are an important element in the malfunctioning of robots, and they occur in robots due to reasons such as embedded software or the controlling software and application software. Some of the methods that can be useful to reduce the occurrence of software faults in robots are failure mode and effects analysis (FMEA), fault tree analysis (FTA), and testing.

4.4 Robot Reliability Measures

There are various types of reliability-related measures associated with robots. Some of these are presented below.

4.4.1 Mean Time to Robot Failure

Mean time to robot failure can be obtained by using either of the following three formulas:

$$MTRF = \int_0^{\infty} R_{rb}(t)\,\mathrm{d}t \tag{4.1}$$

$$MTRF = \lim_{s \to 0} R_{rb}(s) \tag{4.2}$$

$$MTRF = \frac{RPH - DDTRF}{TNRF} \tag{4.3}$$

where

$MTRF$ is the mean time to robot failure.
$R_{rb}(t)$ is the robot reliability at time t.
s is the Laplace transform variable.
$R_{rb}(s)$ is the Laplace transform of the robot reliability function.
$TNRF$ is the total number of robot failures.
RPH is the robot production hours.
$DDTRF$ is the downtime due to robot failure expressed in hours.

Example 4.1
The annual total robot production hours and total downtime due to robot failures in an organization are 60,000 hours and 500 hours, respectively. During the period a total of 20 robot failures has occurred. Calculate the mean time to robot failure.

By substituting the given data values into Equation (4.3), we get

$$MTRF = \frac{60,000 - 500}{20} = 2,975 \text{ hours}$$

Thus, the mean time to robot failure is 2,975 hours.

Example 4.2
Assume that the failure rate, λ_{rb}, of a robot is 0.0005 failures per hour and its reliability is expressed by

$$\begin{aligned} R_{rb}(t) &= e^{-\lambda_{rb} t} \\ &= e^{-(0.0005)t} \end{aligned} \tag{4.4}$$

where

$R_{rb}(t)$ is the robot reliability at time t.

Calculate mean time to robot failure by using Equations (4.1) and (4.2). Comment on the end result.

By substituting Equation (4.4) into Equation (4.1), we obtain

$$MTRF = \int_0^\infty e^{-(0.0005)t} dt = \frac{1}{0.0005} = 2{,}000 \text{ hours}$$

By taking the Laplace transform of Equation (4.4), we get

$$R_{rb}(s) = \frac{1}{(s + 0.0005)} \tag{4.5}$$

where

$R_{rb}(s)$ is the Laplace transform of the reliability function.

Using Equation (4.5) in Equation (4.2) yields

$$\begin{aligned} MTRF &= \lim_{s \to 0} \frac{1}{(s + 0.0005)} \\ &= \frac{1}{0.0005} \\ &= 2{,}000 \text{ hours} \end{aligned}$$

In both cases, the end result (*i. e.*, MTRF = 2,000 hours) is exactly the same.

4.4.2 Mean Time to Robot-related Problems

Mean time to robot-related problems is the average productive robot time prior to the occurrence of a robot-related problem and is defined by

$$MTRP = \frac{RPH - DDTRP}{TNRP} \tag{4.5}$$

where

$MTRP$ is the mean time to robot-related problems.

$DDTRP$ is the downtime due to robot-related problems expressed in hours.

$TNRP$ is the total number of robot-related problems.

Example 4.3

Assume that the annual total robot production hours and total downtime due to robot-related problems, at an industrial installation, are 80,000 hours and

1,000 hours, respectively. During the period a total of 30 robot-related problems has occurred. Calculate the mean time to robot-related problems.

By inserting the specified data values into Equation (4.5) yields

$$\begin{aligned} MTRP &= \frac{80,000 - 1,000}{30} \\ &= 2633.3 \text{ hours} \end{aligned}$$

Thus, the mean time to robot-related problems is 2633.3 hours.

4.4.3 Robot Reliability

This is defined by [6]

$$R_{\text{rb}}(t) = \exp\left[-\int_0^t \lambda_{rb}(t)\,\mathrm{d}t\right] \tag{4.6}$$

where

$\lambda_{\text{rb}}(t)$ is the robot hazard rate or time-dependent failure rate.

Equation (4.6) is the general expression for obtaining robot reliability. More specifically, it can be used to obtain a robot's reliability when robot times to failure follow any statistical distribution (*e. g.*, Weibull, normal, gamma, or exponential).

Example 4.4

A robot's hazard rate is defined by the following function:

$$\lambda_{\text{rb}}(t) \equiv \frac{\theta t^{\theta-1}}{\alpha\alpha^{\theta-1}} \tag{4.7}$$

where

$\lambda_{\text{rb}}(t)$ is the robot's hazard rate when its times to failure follow Weibull distribution.

t is time.

θ is the shape parameter.

α is the scale parameter.

Obtain an expression for the robot reliability and then use the expression to calculate reliability when $t = 500$ hours, $\theta = 1$, and $\alpha = 1,000$ hours.

Using Equation (4.7) in Equation (4.6) yields

$$R_{rb}(t) = \exp\left[-\int_0^t \frac{\theta t^{\theta-1}}{\alpha \alpha^{\theta-1}} \mathrm{d}t\right] = \mathrm{e}^{-\left(\frac{t}{\alpha}\right)^\theta} \tag{4.8}$$

By substituting the given data values into Equation (4.8), we get

$$R_{rb}(500) = \mathrm{e}^{-\left(\frac{500}{1000}\right)} = 0.6065$$

Thus, the robot reliability for the specified mission period of 500 hours is 0.6065.

4.4.4 Robot Hazard Rate

This is defined by [6]:

$$\lambda_{rb}(t) = -\frac{1}{R_{rb}(t)} \cdot \frac{\mathrm{d}R_{rb}(t)}{\mathrm{d}t} \tag{4.9}$$

where

$\lambda_{rb}(t)$ is the robot hazard rate.
$R_{rb}(t)$ is the robot reliability at time t.

Equation (4.9) can be used to obtain robot hazard rate when robot times to failure follow any time-continuous distribution (*e.g.*, exponential, Rayleigh, Weibull, etc.).

Example 4.5
By using Equations (4.8) and (4.9), prove that the robot hazard rate is given by Equation (4.7).

Using Equation (4.8) in Equation (4.9) yields

$$\lambda_{rb}(t) = -\frac{1}{\mathrm{e}^{-\left(\frac{t}{\theta}\right)^\theta}}\left[-\frac{\theta}{\alpha}\left(\frac{t}{\alpha}\right)^{\theta-1}\right]\mathrm{e}^{-\left(\frac{t}{\alpha}\right)^\theta} = \frac{\theta}{\alpha}\frac{t^{\theta-1}}{\alpha^{\theta-1}} \tag{4.10}$$

Both Equations (4.7) and (4.10) are identical. It proves that Equation (4.7) is an expression for robot hazard rate.

4.5 Robot Reliability Analysis Methods

There are many methods used to perform various types of reliability analysis in the field of reliability engineering. Some of them can be used quite effectively to conduct robot reliability-related studies. Four of these methods are shown in Figure 4.2. These are parts count method, failure modes and effect analysis (FMEA), fault tree analysis, and Markov method.

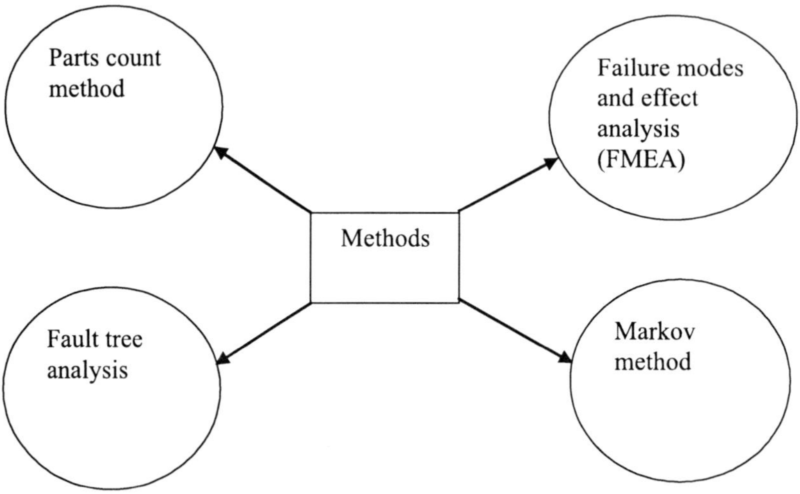

Figure 4.2. Methods for performing robot reliability-related studies

The parts count method is used during bid proposal and early design phases for estimating equipment failure rate. The method requires information on items such as equipment/product use environment, generic part types and quantities, and part quality levels [18].

Additional information on parts count method is available in Refs. [18, 19].

Failure modes and effect analysis (FMEA) is an effective tool to conduct analysis of each failure mode in the system/equipment to determine the effects of such failure modes on the total system/equipment [20]. This method was developed by the United States Department of Defense in the early 1950s and comprises of the following six steps [19–22]:

- Define system/equipment and its associated requirements.
- Develop appropriate ground rules.
- Describe the system/equipment and all its related functional blocks.
- Highlight all possible failure modes and their effects.
- Develop critical items list.
- Document the analysis.

FMEA is described in detail in Chapter 3 and in [19, 23].

Fault tree analysis is a widely used method to evaluate reliability of engineering systems during their design and development phase. A fault tree may be described as a logical representation of the relationship of basic or primary events that result in a specified undesirable event called the "top event". This method was developed in the early 1960s at Bell Telephone Laboratories and is described in detail in Chapter 3 and in [19, 24].

The Markov method can be used in more cases than any other reliability evaluation method and is used to model systems with constant failure and repair rates. The method is described in detail in Chapter 3 and in [19, 25]. Its application to robot-related problems is demonstrated by two of the mathematical models presented in Section 4.6.

4.6 Models for Performing Robot Reliability and Maintenance Studies

There are many mathematical models that can be used to perform various types of robot reliability and maintenance studies. Four of these models are presented below.

4.6.1 Model I

This model represents a robot system that can fail either due to a human error or other failures (*e. g.*, hardware and software). The failed robot system is repaired to its operating state. The robot system state diagram is shown in Figure 4.3. The numerals in the rectangle, circle, and diamond denote system states. The following assumptions are associated with this robot system model [6]:

- Human error and other failures are statistically independent.
- Human error and other failure rates are constant.
- The failed robot system repair rates are constant.
- The repaired robot system is as good as new.

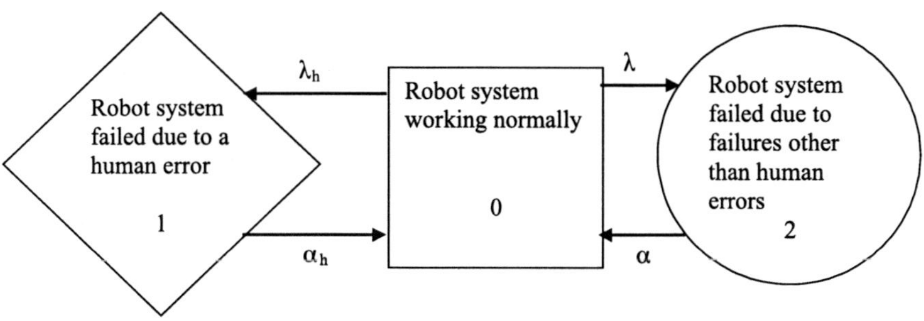

Figure 4.3. Robot system state space diagram

The following symbols are associated with the diagram in Figure 4.2 and its associated equations:

$P_i(t)$ is the probability that the robot system is in state i at time t; for $i=0$ (working normally), $i=1$ (failed due to a human error), $i=2$ (failed due to failures other than human errors).
λ_h is the robot system human error rate.
α_h is the robot system repair rate from failed state 1.
λ is the robot system non-human error failure rate.
α is the robot system repair rate from failed state 2.

Using the Markov method, we write down the following equations for the Figure 4.2 diagram [6]:

$$\frac{dP_0(t)}{dt}+(\lambda+\lambda_h)P_0(t)=\alpha_h P_1(t)+\alpha P_2(t) \tag{4.11}$$

$$\frac{dP_1(t)}{dt}+\alpha_h P_1(t)=\lambda_h P_0(t) \tag{4.12}$$

$$\frac{dP_2(t)}{dt}+\alpha P_2(t)=\lambda P_0(t) \tag{4.13}$$

At time $t=0$, $P_0(0)=1$, $P_1(0)=0$, and $P_2(0)=0$.

Solving Equations (4.11)–(4.13) using Laplace transforms, we get

$$P_0(t)=\frac{\alpha\alpha_h}{m_1 m_2}+\left[\frac{(m_1+\alpha)(m_1+\alpha_h)}{m_1(m_1-m_2)}\right]e^{m_1 t}-\left[\frac{(m_2+\alpha)(m_2+\alpha_h)}{m_2(m_1-m_2)}\right]e^{m_2 t} \tag{4.14}$$

where

$$m_1, m_2=\frac{-b\pm\left[b^2-4(\alpha\alpha_h+\lambda_h\alpha+\lambda\alpha_h)\right]^{1/2}}{2} \tag{4.15}$$

$$b=\lambda+\lambda_h+\alpha+\alpha_h \tag{4.16}$$

$$m_1 m_2=\alpha\alpha_h+\lambda_h\alpha+\lambda\alpha_h \tag{4.17}$$

$$m_1+m_2=(\lambda+\lambda_h+\alpha+\alpha_h) \tag{4.18}$$

$$P_1(t)=\frac{\alpha\lambda_h}{m_1 m_2}+\left[\frac{\lambda_h m_1+\lambda_h\alpha}{m_1(m_1-m_2)}\right]e^{m_1 t}-\left[\frac{(\alpha+m_2)\lambda_h}{m_2(m_1-m_2)}\right]e^{m_2 t} \tag{4.19}$$

$$P_2(t)=\frac{\lambda\alpha_h}{m_1 m_2}+\left[\frac{\lambda m_1+\lambda\alpha_h}{m_1(m_1-m_2)}\right]e^{m_1 t}-\left[\frac{(\alpha_h+m_2)\lambda}{m_2(m_1-m_2)}\right]e^{m_2 t} \tag{4.20}$$

The robot system availability, $AV_r(t)$, is given by

$$AV_{rb}(t) = P_0(t) \tag{4.21}$$

As time t becomes large in Equations (4.19)–(4.21), we get the following steady state probability expressions:

$$AV_{rb} = \frac{\alpha\alpha_h}{m_1 m_2} \tag{4.22}$$

$$P_1 = \frac{\alpha\lambda_h}{m_1 m_2} \tag{4.23}$$

$$P_2 = \frac{\lambda\alpha_h}{m_1 m_2} \tag{4.24}$$

where

AV_{rb} is the robot system steady state availability.
P_1 is the steady state probability of the robot system being in state 1.
P_2 is the steady state probability of the robot system being in state 2.

For $\alpha = \alpha_h = 0$, from Equations (4.14), (4.19), and (4.20) we get

$$P_0(t) = e^{-(\lambda+\lambda_h)t} \tag{4.25}$$

$$P_1(t) = \frac{\lambda_h}{(\lambda+\lambda_h)}\left[1 - e^{-(\lambda+\lambda_h)t}\right] \tag{4.26}$$

$$P_2(t) = \frac{\lambda}{(\lambda+\lambda_h)}\left[1 - e^{-(\lambda+\lambda_h)t}\right] \tag{4.27}$$

The robot system reliability at time t from Equation (4.25) is

$$R_{rb}(t) = e^{-(\lambda+\lambda_h)t} \tag{4.28}$$

where

$R_{rb}(t)$ is the robot system reliability at time t.

By substituting Equation (4.28) into Equation (4.1), we get the following expression for mean time to robot failure:

$$\begin{aligned} MTRF &= \int_0^\infty e^{-(\lambda+\lambda_h)t}\,dt \\ &= \frac{1}{\lambda+\lambda_h} \end{aligned} \tag{4.29}$$

Using Equation (4.28) in Equation (4.9) yields the following expression for the robot system hazard rate:

$$\begin{aligned}\lambda_{rb}(t) &= -\frac{1}{e^{-(\lambda+\lambda_h)t}} \cdot \frac{d\left[e^{-(\lambda+\lambda_h)t}\right]}{dt} \\ &= \lambda + \lambda_h\end{aligned} \tag{4.30}$$

The right-hand side of Equation (4.30) is independent of time, which means that the robot system failure rate is constant.

Example 4.6

A robot system can fail either due to human error or other failures, and its constant human error and other failure rates are 0.0004 errors per hour and 0.0008 failures per hour, respectively. The robot system constant repair rate from both the failure modes is 0.009 repairs per hour. Calculate the robot system steady state availability by using Equations (4.23) and (4.24).

By substituting the specified data values into Equations (4.23) and (4.24), we get

$$\begin{aligned}P_1 &= \frac{(0.009)(0.0004)}{(0.009)(0.009)+(0.0004)(0.009)+(0.0008)(0.009)} \\ &= 0.0392\end{aligned}$$

and

$$\begin{aligned}P_2 &= \frac{(0.0008)(0.009)}{(0.009)(0.009)+(0.0004)(0.009)+(0.0008)(0.009)} \\ &= 0.0784\end{aligned}$$

Thus, the robot system steady state unavailability, UAV_{rb}, is

$$\begin{aligned}UAV_{rb} &= P_1 + P_2 = 0.0392 + 0.0784 \\ &= 0.1176\end{aligned}$$

The robot system steady state availability, AV_{rb}, by using the above-calculated value is

$$\begin{aligned}AV_{rb} &= 1 - UAV_{rb} \\ &= 1 - 0.1176 \\ &= 0.8824\end{aligned}$$

4.6.2 Model II

This model is concerned with determining the economic life of a robot, more specifically, the time limit beyond which it is not economical to carry out repairs. Thus, the economic life, T_e, of the robot is expressed as [26–28]:

$$T_e = \left[\frac{2(C_{ir} - SV_r)}{C_{rin}}\right]^{1/2} \tag{4.31}$$

where

C_{rin} is the annual increase in robot repair cost.
SV_r is the robot scrap value.
C_{ir} is the robot initial cost (installed).

Example 4.7
Assume that a robot costs \$90,000 (installed) and its estimated scrap value is \$2,000. The estimated annual increase in repair cost is \$500. Calculate the time limit beyond which the robot repairs will not be beneficial.

Inserting the given data values into Equation (4.31) yields

$$T_e = \left[\frac{2(90{,}000 - 2{,}000)}{500}\right]^{1/2}$$
$$= 18.76 \text{ years}$$

Thus, the time limit beyond which the robot repairs will not be economical or beneficial is 18.76 years.

4.6.3 Model III

This model can be used to calculate the optimum number of inspections per robot facility per unit time [28]. This information is useful to decision makers because inspections are often disruptive; however, such inspections usually reduce the robot downtime because they lead to fewer breakdowns. In this model, the total robot downtime is minimized to get the optimum number of inspections.

The total robot downtime, *TRDT*, per unit time is defined as [29]

$$TRDT = nT_{dp} + \frac{kT_{db}}{n} \tag{4.32}$$

where

n is the number of inspections per robot facility per unit time.
T_{dp} is the downtime per inspection for a robot facility.
T_{db} is the downtime per breakdown for a robot facility.
k is a constant for a specific robot facility.

By differentiating Equation (4.32) with respect to n and then equating it to zero, we get

$$n^* = \left[\frac{kT_{\text{db}}}{T_{\text{dp}}}\right]^{1/2} \quad (4.33)$$

where

n^* is the optimum number of inspections per robot facility per unit time.

By substituting Equation (4.33) into Equation (4.32), we get

$$TRDT^* = 2\left[T_{\text{dp}} \cdot kT_{\text{db}}\right]^{1/2} \quad (4.34)$$

where

$TRDT^*$ is the minimum total robot downtime.

Example 4.8

Assume that for a certain robot facility, the following data are specified:

- $T_{\text{dp}} = 0.04$ months
- $T_{\text{db}} = 0.12$ months
- $k = 3$

Compute the optimum number of robot inspections per month and the minimum total robot downtime.

By substituting the above given values into Equations (4.33) and (4.34), we get

$$n^* = \left[\frac{3(0.12)}{0.04}\right]^{1/2} = 3 \text{ inspections per month}$$

and

$$TRDT^* = 2\left[(0.04)(3)(0.12)\right]^{1/2}$$
$$= 0.24 \text{ months}$$

4.6.4 Model IV

This model represents a robot system composed of a robot and a safety unit. In the industrial sector, the inclusion of safety units or systems with robots is often practiced because of robot accidents involving humans. In this model, it is assumed that after the failure of the safety unit, the robot may fail safely or with an incident. The failed safety unit is repaired. The robot system state space diagram is shown in Figure 4.4. The numerals in boxes and circles denote system states.

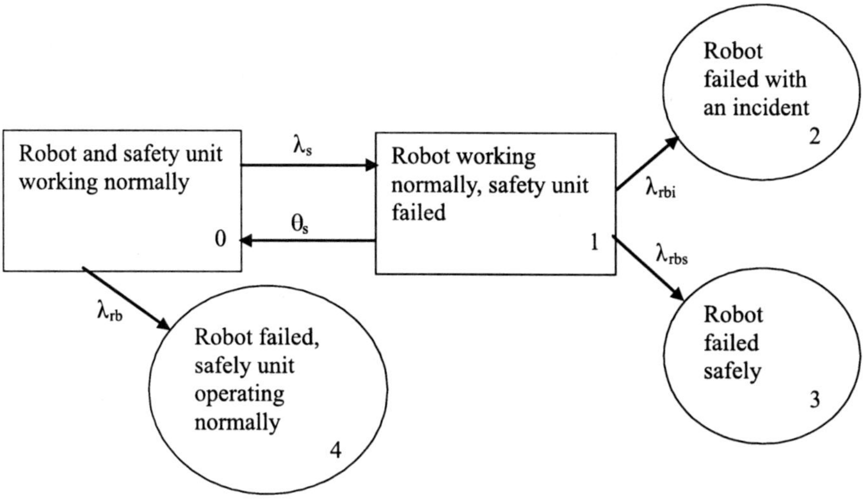

Figure 4.4. Robot system state space diagram

The following assumptions are associated with this model:

- All failures are statistically independent.
- All failure and repair rates are constant.
- The robot system fails when the robot fails.
- The repaired safety unit is as good as new.

The following symbols are associated with the diagram in Figure 4.4 and its associated equations:

$P_i(t)$ is the probability that the robot system is in state i at time t; for $i=0$ (robot and safety unit working normally), $i=1$ (robot working normally, safety unit failed), $i=2$ (robot failed with an incident), $i=3$ (robot failed safely), $i=4$ (robot failed, safety unit operating normally).
λ_{rb} is the robot failure rate.
λ_s is the safety unit failure rate.
λ_{rbi} is the rate of the robot failing with an incident.
λ_{rbs} is the rate of the robot failing safely.
θ_s is the safety unit repair rate.

Using the Markov method, we write down the following equations for the diagram in Figure 4.3 [30]:

$$\frac{dP_0(t)}{dt} + (\lambda_{rb} + \lambda_s) P_0(t) = \theta_s P_1(t) \tag{4.35}$$

$$\frac{dP_1(t)}{dt} + (\lambda_{rbi} + \lambda_{rbs} + \theta_s) P_1(t) = \lambda_s P_0(t) \tag{4.36}$$

$$\frac{\mathrm{d}P_2(t)}{\mathrm{d}t} = \lambda_{\mathrm{rbi}} P_1(t) \tag{4.37}$$

$$\frac{\mathrm{d}P_3(t)}{\mathrm{d}t} = \lambda_{\mathrm{rbs}} P_1(t) \tag{4.38}$$

$$\frac{\mathrm{d}P_4(t)}{\mathrm{d}t} = \lambda_{\mathrm{rb}} P_0(t) \tag{4.39}$$

At time $t=0$, $P_0(0)=1$, $P_1(0)=0$, $P_2(0)=0$, $P_3(0)=0$, and $P_4(0)=0$.

Solving Equations (4.35)–(4.39) using Laplace transforms, we get

$$P_0(t) = \mathrm{e}^{-At} + \lambda_s \theta_s \left[\frac{\mathrm{e}^{-At}}{(c_1+A)(c_2+A)} + \frac{\mathrm{e}^{c_1 t}}{(c_1+A)(c_2-c_2)} + \frac{\mathrm{e}^{c_2 t}}{(c_2+A)(c_2-c_1)} \right] \tag{4.40}$$

where

$$A = \lambda_{\mathrm{s}} + \lambda_{\mathrm{rb}} \tag{4.41}$$

$$c_1, c_2 = \frac{-B \pm \left(B^2 - 4F\right)^{1/2}}{2} \tag{4.42}$$

$$B = A + \theta_{\mathrm{s}} + \lambda_{\mathrm{rbi}} + \lambda_{\mathrm{rbs}} \tag{4.43}$$

$$F = \lambda_{\mathrm{rbi}} \lambda_{\mathrm{s}} + \lambda_{\mathrm{rbs}} \lambda_{\mathrm{s}} + \lambda_{\mathrm{rbi}} \lambda_{\mathrm{rb}} + \lambda_{\mathrm{rbs}} \lambda_{\mathrm{rb}} + \theta_{\mathrm{s}} \lambda_{\mathrm{rb}} \tag{4.44}$$

$$P_1(t) = \lambda_{\mathrm{s}} \left[\left(\mathrm{e}^{c_1 t} - \mathrm{e}^{c_2 t} \right) / \left(c_1 - c_2 \right) \right] \tag{4.45}$$

$$P_2(t) = \frac{\lambda_{\mathrm{rbi}} \lambda_{\mathrm{s}}}{c_1 c_2} \left[1 + \left(c_1 \mathrm{e}^{c_2 t} - c_2 \mathrm{e}^{c_1 t} \right) / \left(c_2 - c_1 \right) \right] \tag{4.46}$$

$$P_3(t) = \frac{\lambda_{\mathrm{rbs}} \lambda_{\mathrm{s}}}{c_1 c_2} \left[1 + \left(c_1 \mathrm{e}^{c_2 t} - c_2 \mathrm{e}^{c_1 t} \right) / \left(c_2 - c_1 \right) \right] \tag{4.47}$$

$$P_4(t) = \frac{\lambda_{\mathrm{rb}}}{A}\left(1 - \mathrm{e}^{-At}\right) + \lambda_{\mathrm{s}} \lambda_{\mathrm{rb}} \theta_{\mathrm{s}} \left[\frac{1}{c_1 c_2 A} - \frac{\mathrm{e}^{-At}}{A(c_1+A)(c_2+A)} + \frac{\mathrm{e}^{c_1 t}}{c_1(c_1+A)(c_1-c_2)} + \frac{\mathrm{e}^{c_2 t}}{c_2(c_2+A)(c_2-c_1)} \right] \tag{4.48}$$

The robot system reliability (*i. e.*, when both robot and safety unit work normally) with safety unit repair facility is given by

$$R_{rbr}(t) = P_0(t) \tag{4.49}$$

By substituting Equation (4.49) into Equation (4.1), we get the following expression for robot system mean time to failure:

$$MTTF_{rbr} = \frac{1}{A}\left[1 + \frac{\lambda_s \theta_s}{F}\right] \tag{4.50}$$

where

$MTTF_{rbr}$ is the robot system mean time to failure (*i. e.*, when both robot and safety unit are working) with safety unit repair facility.

Example 4.9

Assume that a robot system is composed of a robot and a safety unit. The operating robot with failed safety unit can either fail with an incident or safely and the failed safety unit is repaired.

Calculate the robot system mean time to failure by using Equation (4.50) for the following given data values:

- $\lambda_{rb} = 0.0005$ failures per hour
- $\lambda_s = 0.0003$ failures per hour
- $\lambda_{rbi} = 0.0002$ failures per hour
- $\lambda_{rbs} = 0.0003$ failures per hour
- $\theta_s = 0.008$ repairs per hour

Using the above data values in Equation (4.50) yields

$$MTTF = \frac{1}{A}\left[1 + \frac{(0.0003)(0.008)}{F}\right]$$

where

$A = 0.0003 + 0.0005 = 0.0008$

$$F = (0.0002)(0.0003) + (0.0003)^2 + (0.0002)(0.0005)$$
$$+ (0.0003)(0.0005) + (0.008)(0.0005)$$
$$= 0.0000044$$

$$MTTF = \frac{1}{(0.0008)}\left[1 + \frac{(0.0003)(0.008)}{(0.0000044)}\right]$$
$$= 1931.8 \text{ hours}$$

Thus, the robot system mean time to failure is 1931.8 hours.

4.7 Problems

1. Write an essay on historical developments in robot reliability.
2. Define the following terms:
   - Robot reliability
   - Graceful failure
   - Fail-safe
3. List at least six common causes of robot failures.
4. Write down three formulas that can be used to calculate mean time to robot failure.
5. Assume that the reliability of a robot is defined by the following equation:

$$R_{rb}(t) = e^{-0.002t} \tag{4.51}$$

   where
   $R_{rb}(t)$ is the robot reliability at time t.
6. Prove Equation (4.7) by using Equation (4.8).
7. Discuss the parts count method.
8. Discuss at least three methods that can be used to perform robot reliability analysis.
9. Prove Equations (4.22)–(4.25).
10. A robot costs $100,000 (installed) and its estimated scrap value is $4,000. The estimated annual increase in repair cost is $600. Calculate the time limit beyond which the robot repairs will not be beneficial.

References

1 Jablonowski, J., Posey, J.W., Robotics Terminology, in Handbook of Industrial Robotics, edited by S.Y. Nof, John Wiley and Sons, New York, 1985, pp. 1271–1303.
2 Zeldman, M.I., What Every Engineer Should Know About Robots, Marcel Dekker, New York, 1984.
3 Rudall, B.H., Automation and Robotics Worldwide: Reports and Surveys, Robotica, Vol. 14, 1996, pp. 164–168.
4 Engleberger, J.F., Three Million Hours of Robot Field Experience, The Industrial Robot, 1974, pp. 164–168.
5 Dhillon, B.S., On Robot Reliability and Safety: Bibliography, Microelectronics and Reliability, Vol. 27, 1987, pp. 105–118.
6 Dhillon, B.S., Robot Reliability and Safety, Springer-Verlag, New York, 1991.
7 Dhillon, B.S., Fashandi, A.R.M., Liu, K.L., Robot Systems Reliability and Safety: A Review, Journal of Quality in Maintenance Engineering, Vol. 8, No. 3, 2002, pp. 170–212.

8 Jones, R., Dawson, S., People and Robots: Their Safety and Reliability, Proceedings of the 7th British Robot Association Annual Conference, 1984, pp. 243–258.
9 American National Standard for Industrial Robots and Robot Systems: Safety Requirements, ANSI/RIA R15.06–1986, American National Standards Institute (ANSI), New York, 1986.
10 Glossary of Robotics Terminology, in Robotics, edited by E.L. Fisher, Industrial Engineering and Management Press, Institute of Industrial Engineers, Atlanta, Georgia, 1983, pp. 231–253.
11 Tver, D.F., Bolz, R.W., Robotics Sourcebook and Dictionary, Industrial Press, New York, 1983.
12 Susnjara, K.A., A Manager's Guide to Industrial Robots, Corinthian Press, Shaker Heights, Ohio, 1982.
13 Sato, K., Case Study of Maintenance of Spot-Welding Robots, Plant Maintenance, Vol. 14, 1982, pp. 28–29.
14 Sugimoto, N., Kawaguchi, K., Fault Tree Analysis of Hazards Created by Robots, Proceedings of the 13th International Symposium on Industrial Robots, 1983, pp. 9.13–9.28.
15 Khodabandehloo, K., Duggan, F., Husband, T.F., Reliability Assessment of Industrial Robots, Proceedings of the 14th International Symposium on Industrial Robots, 1984, pp. 209–220.
16 Khodabandehloo, K., Duggan, F., Husband, R.G., Reliability of Industrial Robots: A Safety Viewpoint, Proceedings of the 7th British Robot Association Annual Conference, 1984, pp. 233–242.
17 Dhillon, B.S., Human Reliability: with Human Factors, Pergamon Press, New York, 1986.
18 MIL-HDBK-217, Reliability Prediction of Electronic Equipment, Department of Defense, Washington, D.C.
19 Dhillon, B.S., Design Reliability: Fundamentals and Applications, CRC Press, Boca Raton, Florida, 1999.
20 Omdahl, T.P., Editor, Reliability, Availability, and Maintainability (RAM) Dictionary, American Society for Quality Control (ASQC) Press, Milwaukee, Wisconsin, 1988.
21 MIL-F-18372 (Aer), General Specification for Design, Installation, and Test of Aircraft Flight Control Systems, Bureau of Naval Weapons, Department of the Navy, Washington, D.C., Para. 3.5.2.3.
22 Coutinho, J.S., Failure Effect Analysis, Trans. New York Academy of Sciences, Vol. 26, Series II, 1963–1964, pp. 564–584.
23 Palady, P., Failure Modes and Effects Analysis, PT Publications, West Palm Beach, Florida, 1995.
24 Fault Tree Handbook, Report No. NUREG-0492, U.S. Nuclear Regulatory Commission, Washington, D.C., 1981.
25 Shooman, M.L., Probabilistic Reliability: An Engineering Approach, McGraw Hill Book Company, New York, 1968.
26 Varnum, E.C., Bassett, B.B., Machine and Tool Replacement Practices, in Manufacturing Planning and Estimating Handbook, edited by F.W. Wilson and P.D. Harvey, McGraw Hill Book Company, New York, 1963, pp. 18.1–18.22.
27 Eidmann, F.L., Economic Control of Engineering and Manufacturing, McGraw Hill Book Company, New York, 1931.

28 Dhillon, B.S., Mechanical Reliability: Theory, Models, and Applications, American Institute of Aeronautics and Astronautics, Washington, D.C., 1988.
29 Wild, R., Essentials of Production and Operations Management, Holt, Reinhart, and Winston, London, 1985, pp. 356–368.
30 Dhillon, B.S., Yang, N., Reliability Analysis of a Repairable Robot System, Journal of Quality in Maintenance Engineering, Vol. 2, 1996, pp. 30–37.

5

Medical Equipment Reliability

5.1 Introduction

The history of the earliest use of medical devices may be traced back to the ancient Egyptians and Etruscans using various types of dental devices [1]. Today medical devices and equipment are widely used throughout the world. In fact, in 1988 the world medical equipment production was estimated to be around $36 billion [1], and in 1997, the world market for medical devices was valued at around $120 billion [2].

The beginning of the medical equipment or device reliability field may be traced back to the latter part of the 1960s, when a number of publications on the subject appeared [3–7]. These publications covered topics such as "Instrument induced errors in the electrocardiogram", "Reliability of ECG instrumentation", "Safety and reliability in medical electronics", and "The effect of medical test instrument reliability on patient risks" [3–6]. In 1980, an article presented a comprehensive list of publications on medical equipment reliability [8], and in 1983 a text on reliability engineering devoted one entire chapter to medical/equipment reliability [9]. In 2000, a book entitled *Medical Device Reliability and Associated Areas* provided a comprehensive list of publications on the subject [10]. More recent publications on medical device/equipment reliability are listed at the end of this book. This chapter presents various important aspects of medical equipment reliability.

5.2 Medical Equipment Reliability-related Facts and Figures

Some of the facts and figures directly or indirectly related to medical equipment/device reliability are as follows:

- In 1997, there were a total of 10,420 registered medical device manufacturers in the United States [11].
- Due to faulty medical instrumentation around 1,200 deaths per year occur in the United States [12, 13].
- In 1969, the United States Department of Health, Education, and Welfare special committee reported that over a 10 year period, around 10,000 injuries were associated with medical devices/equipment and 731 resulted in deaths [14, 15].
- A study reported that over 50% of all technical medical equipment problems were due to operator errors [16].
- A study reported that around 100,000 Americans die each year due to human errors, and their financial impact on the United States economy was estimated to be between $17 billion and $29 billion [17].
- The Emergency Care Research Institute (ECRI) tested a sample of 15,000 products used in hospitals and found that around 4% to 6% of these products were sufficiently dangerous to warrant immediate correction [16].
- In 1990, a study performed by the US Food and Drug Administration (FDA) revealed that around 44% of the quality-related problems that resulted in the voluntary recall of medical devices for the period October 1983 to September 1989 were the result of deficiencies/errors that could have been prevented through effective design controls [18].

5.3 Medical Devices and Classification of Medical Devices/Equipment

Today, there are over 5,000 different types of medical devices being used in a modern hospital. They range from a simple tongue depressor to a complex pacemaker [1, 10]. Thus, the criticality of their reliability varies from one device to another. Nonetheless, past experiences indicate that the failure of medical devices has been very costly in terms of fatalities, injuries, dollar and cents, etc. Needless to say, modern medical devices and equipment have become very complex and sophisticated and are expected to operate under stringent environments.

Electronic equipment used in the health care system may be classified under the following three categories [7]:

- **Category I.** This category includes those medical equipment/devices that are directly and immediately responsible for the patient's life or may become so under emergency conditions. When such equipment fails, there is seldom sufficient time for the repair action. Thus, this type of equipment must always operate successfully at the moment of need. Some examples of such equipment/devices are as follows:
  - Respirators
  - Cardiac pacemakers
  - Electrocardiographic monitors
  - Cardiac defibrillators

- **Category II.** This category includes those medical equipment/devices that are used for routine or semi-emergency diagnostic or therapeutic purposes. Failure of such equipment or devices is not as critical as those fall under Category I, because there is time for repair. Some examples of such equipment/devices are as follows:
  - Spectrophotometers
  - Gas analyzers
  - Electrocardiograph and electroencephalograph recorders and monitors
  - Diathermy equipment
  - Ultrasound equipment
  - Colorimeters
- **Category III.** This category includes those equipment/devices that are not critical to a patient's life or welfare but serve as convenience equipment or devices. Three examples of such equipment or devices are as follows:
  - Electric beds
  - Wheel chairs
  - Bedside television sets

All in all, there could be some overlap between the above three categories of equipment, particularly between categories I and II. An electrocardiograph recorder or monitor is a typical example of such equipment.

5.4 Medical Equipment Reliability Improvement Procedures and Methods

There are many procedures and methods used to improve medical equipment reliability. Some of these are presented below.

5.4.1 General Approach

The general approach is a 13-step approach developed by Bio-Optronics to produce safe and reliable medical devices [19]. The approach steps are shown in Figure 5.1.

5.4.2 Parts Count Method

The parts count method is used to predict equipment or system failure during the bid proposal and early design stages [20]. The method requires information on areas shown in Figure 5.2.

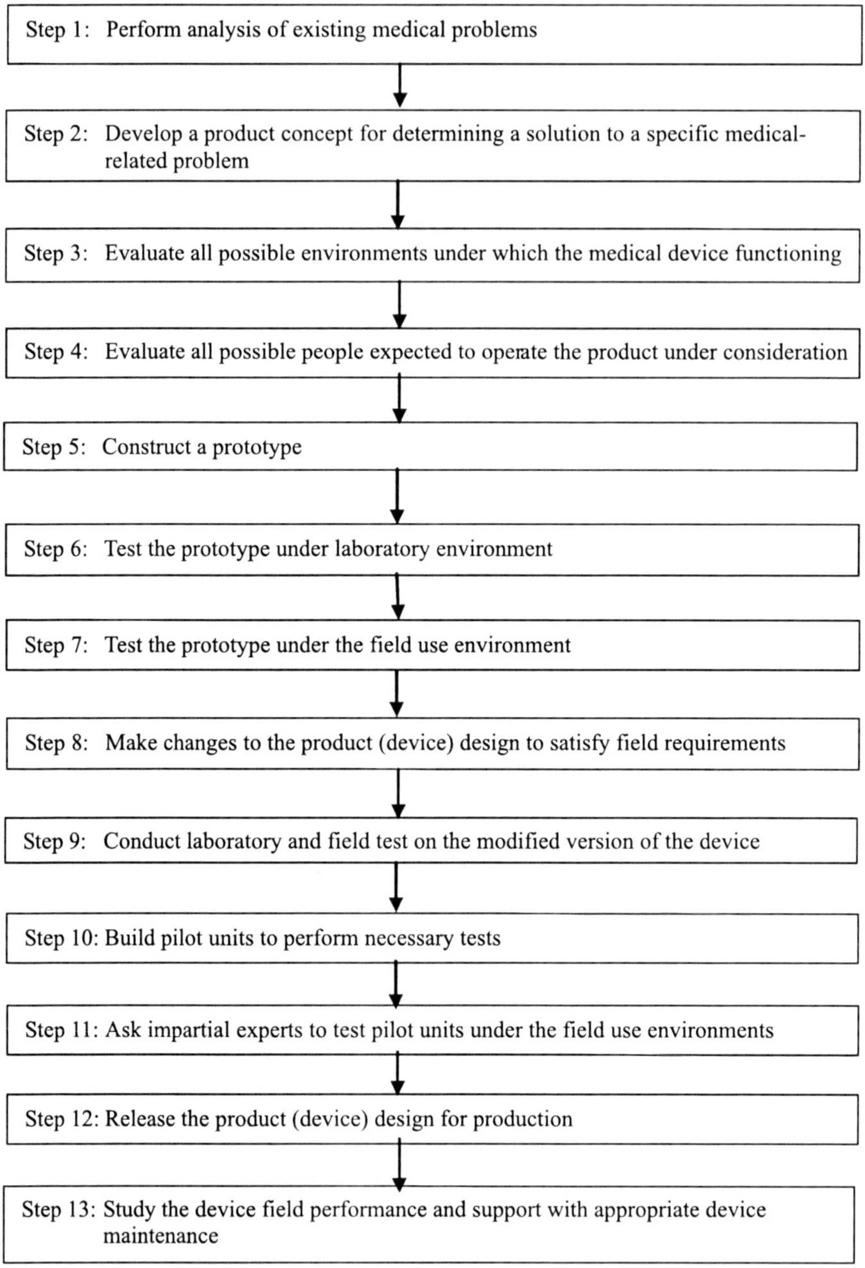

Figure 5.1. An approach for producing safe and reliable medical devices

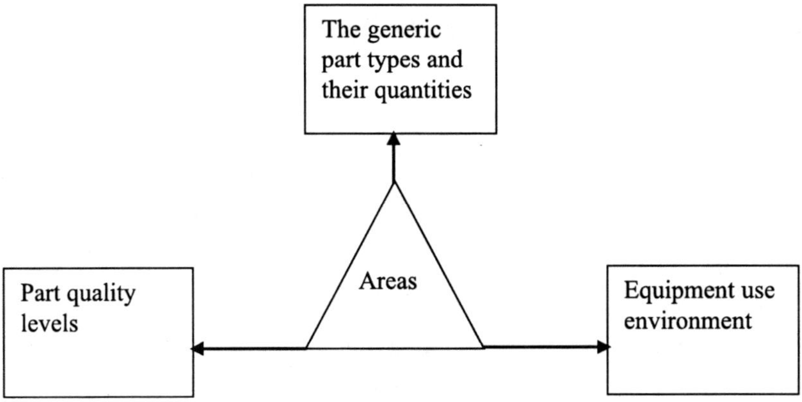

Figure 5.2. Areas of information required by the parts count method

The method calculates the system or equipment failure rate under the single-use environment by using the following equation [20]:

$$\lambda_S = \sum_{i=1}^{n} \theta_i \left(\lambda_g Q_g\right)_i \tag{5.1}$$

where

λ_S is the system failure rate expressed in failures/10^6 hours.
n is the number of different generic component classifications.
Q_g is the generic component quality factor.
λ_g is the generic part failure rate expressed in failures/10^6 hours.
θ_i the generic part quantity for classification i.

The values of Q_g and λ_g are tabulated in Ref. [20], and additional information on the method is available in Refs. [20,21].

5.4.3 Markov Method

The markov method is a very general approach and can generally handle more cases than any other method or technique. It can be used in situations when the components or parts are independent as well as for equipment/systems involving dependent failure and repair modes.

The method proceeds by the enumeration of system states. The state probabilities are then computed, and the steady-state reliability measures can be calculated by applying the frequency balancing method [22]. Additional information on this method is available in Chapter 3 and in Refs. [23,24].

5.4.4 Failure Mode and Effect Analysis (FMEA)

Failure mode and effect analysis (FMEA) is a widely used tool to evaluate design at the early stage from the reliability aspect. This criterion is extremely useful to identify the need for and the effects of design change. FMEA requires the listing of all possible failure modes of each component on paper and their effects on the listed subsystems, etc. The method is known as failure modes, effects, and criticality analysis (FMECA) when criticalities or priorities are assigned to failure mode effects.

Some of the important characteristics of FMEA are as follows [25]:

- It is an upward approach that starts at the detailed level.
- By examining failure effects of all components, the entire system is screened completely.
- It is an effective tool to identify weak spots in system design and indicate areas where further or detailed analysis is desirable.

Additional information on FMEA is available in Chapter 3 and in Refs. [25–26].

5.4.5 Fault Tree Analysis

Fault tree analysis (FTA) begins by identifying an undesirable event, called the top event, associated with a system under consideration [27]. Fault events which could cause the occurrence of the top event are generated and connected by logic operators such as AND and OR. The AND gate provides a TRUE (failed) output when all its inputs are TRUE (failures). In contrast, the OR gate provides a TRUE (failure) output when only one OR more of its inputs are true (failures). All in all, the fault tree construction proceeds by generation of events in a successive manner until the events need not be developed any further.

Additional information on FTA is available in Chapter 3 and in Refs. [27, 28].

5.5 Human Error in Medical Equipment

Human errors are universal and are committed each day. Past experiences indicate that although most are trivial, some can be quite serious or fatal. In the area of health care, one study revealed that in a typical year around 100,000 Americans die due to human errors [17]. Nonetheless, some of the medical equipment/device-related, directly or indirectly, human error facts and figures are as follows:

- The Center for Devices and Radiological Health (CDRH) of the Food and Drug Administration reported that human errors account for 60% of all device-related deaths or injuries in the United States [29].
- Over 50% of all technical medical equipment problems are due to operator errors [16].
- Human error is responsible for up to 90% of accidents both generally and in medical devices [30–31].

- A fatal radiation overdose accident involving the Therac radiation therapy device was the result of a human error [32].
- A patient was seriously injured by over-infusion because the attending nurse incorrectly read the number 7 as 1 [33].

5.5.1 Medical Devices with High Incidence of Human Error

As per Ref. [34], each day human errors in using medical devices cause at least three deaths or serious injuries. Over the years, many studies have been conducted to identify medical devices with a high occurrence of human error. Consequently, the most error-prone medical devices were identified. These devices, in the order of most error-prone to least error-prone, are as follows [34]:

- Glucose meter
- Balloon catheter
- Orthodontic bracket aligner
- Administration kit for peritoneal dialysis
- Permanent pacemaker electrode
- Implantable spinal cord simulator
- Intra-vascular catheter
- Infusion pump
- Urological catheter
- Electrosurgical cutting and coagulation device
- Non-powered suction apparatus
- Mechanical/hydraulic impotence device
- Implantable pacemaker
- Peritoneal dialysate delivery system
- Catheter introducer
- Catheter guide wire
- Trans-luminal coronary angioplasty catheter
- External low-energy defibrillator
- Continuous ventilator (respirator)
- Contact lens cleaning and disinfecting solutions

5.5.2 Important Medical Device/Equipment Operator Errors

There are many types of operator-related errors that occur during medical device/equipment operation or maintenance. Some of the important ones are as follows [35]:

- Incorrect interpretation of or failure to recognize critical device outputs
- Mistakes in setting device parameters

- Incorrect decision-making and actions in critical moments
- Misassembly
- Departure from following specified instructions and procedures
- Inadvertent or untimely activation of controls
- Over-reliance on automatic features of devices/equipment
- Wrong selection of devices in regard to the clinical objectives and requirements

5.6 Useful Guidelines for Reliability and Other Professionals to Improve Medical Equipment Reliability

There is a large number of professionals involved in the manufacture and use of various types of medical devices. Reliability analysts and engineers are one of them. Nonetheless, some of the useful guidelines for reliability and other professionals to improve medical equipment reliability are as follows [24, 36]:

- Reliability professionals
  - Use methods such as FMEA, qualitative FTA, design review, and parts review to obtain immediate results.
  - Focus on critical failures as not all device failures are equally important.
  - Aim to use simple reliability methods as much as possible instead of some sophisticated approaches used in the aerospace industry.
  - Keep in mind that manufacturers are responsible for reliability during the device design and manufacturing phase, and during its operational phase it is basically the responsibility of users.
  - Focus on cost effectiveness and always keep in mind that some reliability improvement decisions require very little or no additional expenditure.
- Other professionals
  - Recognize that failures are the cause of poor medical device reliability, and positive thinking and measures can be quite useful to improve device reliability.
  - For the total success with respect to device reliability, both manufacturers and users must accept their share of related responsibilities.
  - Compare human body and medical device failures. Both of them require appropriate measures from reliability professionals and doctors to enhance device reliability and extend human life, respectively.
  - Remember that the cost of failures is probably the largest single expense in a business organization. Such failures could be associated with equipment, people, business systems, etc., and a reduction in these failures can decrease the cost of business quite significantly.
  - Keep in mind that the application of reliability principles have successfully improved the reliability of systems used in the aerospace area, and their applications to medical devices can generate similar dividends.

5.7 Medical Equipment Maintenance and Maintainability

Medical equipment maintenance may simply be described as all actions necessary for retaining medical equipment in, or restoring to, a specified condition. Similarly, medical equipment maintainability is the probability that a failed piece of medical equipment will be restored to its acceptable operating state. Both these items (*i. e.*, medical equipment maintenance and maintainability) are discussed below, separately [37, 38].

5.7.1 Medical Equipment Maintenance

For the purpose of repair and maintenance, medical equipment may be classified under six classifications: patient diagnostic equipment (*e. g.*, Spiro meters, endoscopes, and physiologic monitors), life support and therapeutic equipment (*e. g.*, ventilators, lasers, and anaesthesia machines), imaging and radiation therapy equipment (*e. g.*, linear accelerators, X-ray machines, and ultrasound devices), laboratory apparatus (*e. g.*, lab analyzers, lab refrigeration equipment, and centrifuges), patient environmental and transport equipment (*e. g.*, patient beds, wheelchairs, and patient-room furniture), and miscellaneous equipment (*e. g.*, all other items that are not included in the other five classifications, for example, sterilizers) [39].

Indices

Just like in the case of the general maintenance activity, there are many indices that can be used to measure the effectiveness of the medical equipment maintenance activity.

Three of these indices are presented below [39].

- Index I

  Index I is a cost ratio and is expressed by

$$\theta_C = \frac{C_s}{C_a} \tag{5.2}$$

where

θ_C is the cost ratio.
C_a is the medical equipment acquisition cost.
C_s is the medical equipment service cost. It includes all labour, parts, and material costs for scheduled and unscheduled service, including in-house, vendor, prepaid contracts, and maintenance insurance.

A range of values for this index, for various classifications of medical equipment, are given in Ref. [10].

- Index II

Index II measures how much time elapses from a customer request until the failed medical equipment is repaired and put back in service. The index is expressed by

$$\theta_{at} = \frac{T_t}{N} \tag{5.3}$$

where
θ_{at} is the average turnaround time per repair.
N is the total number of work orders or repairs.
T_t is the total turnaround time.

As per one study, the turnaround time per medical equipment repair ranged from 35.4 to 135 hours [10].

- Index III

Index III measures how frequently the customer has to request for service per medical equipment. The index is expressed by

$$\theta_C = \frac{R_r}{M} \tag{5.4}$$

where
θ_C is the number of repair requests completed per medical equipment.
R_r is the total number of repair requests.
M is the number of pieces of medical equipment.

As per one study, the value of θ_C ranged from 0.3 to 2 [10].

Mathematical Models

Over the years, a large number of mathematical models concerning engineering equipment maintenance have been developed. Some of these models can equally be used in the area of medical equipment maintenance. One of these models is presented below.

- Model

This model can be used to determine the optimum time interval between item replacements. The model assumes that the item/equipment average annual cost is made up of average investment, operating, and maintenance costs. Thus, the average annual total cost of a piece of equipment is expressed by

$$C_t = C_o + C_m + \frac{C_i}{t} + \frac{(t-1)}{2}[i + j] \tag{5.5}$$

where

C_t is the average annual total cost of a piece of equipment.
i is the amount by which maintenance cost increases annually.
j is the amount by which operational cost increases annually.
C_o is the item/equipment operational cost for the first year.
C_m is the item/equipment maintenance cost for the first year.
C_i is the investment cost.
t is the item/equipment life expressed in years.

Differentiating Equation (5.5) with respect to t and then equating it to zero, yields

$$t^* = \left[\frac{2C_i}{i+j}\right]^{1/2} \tag{5.6}$$

where

t^* is the optimum time between item/equipment replacements.

Example 5.1
Assume that we have following data for a medical equipment:

$i = \$1,000$
$j = \$4,000$
$C_i = \$400,000$

Determine the optimum replacement period for the equipment under consideration.

By inserting the above given data values into Equation (5.6), we get

$$t^* = \left[\frac{2(400,000)}{1000+4000}\right]^{1/2}$$
$$= 12.7 \text{ years}$$

Thus, the optimum replacement period for the medical equipment under consideration is 12.7 years.

5.7.2 Medical Equipment Maintainability

Past experiences indicate that the application of maintainability principles during designing engineering equipment has helped to produce effectively maintainable end products. Their application in the design of medical equipment can also be helpful to produce effectively maintainable end medical items. This section presents three aspects of maintainability considered useful to produce effectively maintainable medical equipment.

Reasons for the Application of Maintainability Principles

Some of the main reasons for applying maintainability principles are as follows [40]:

- To reduce projected maintenance time
- To reduce projected maintenance cost through design modifications
- To determine the number of labour hours and related resources needed to carry out the projected maintenance
- To determine the amount of downtime due to maintenance

Maintainability Design Factors

There are many maintainability design factors. Some of the most frequently addressed factors are shown in Figure 5.3 [41]. Each of these factors is described in detail in Refs. [10, 41].

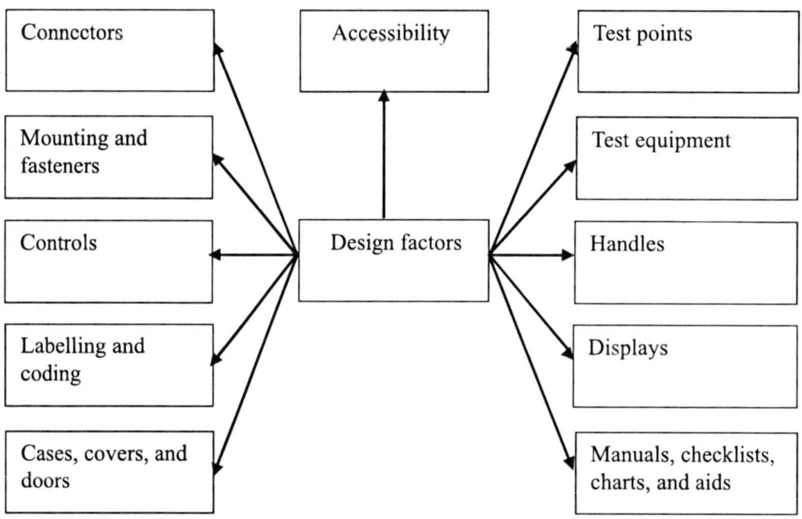

Figure 5.3. Frequently addressed maintainability design factors

Maintainability Measures

There are various types of maintainability measures used in performing maintainability analysis of engineering equipment. Two of these measures are presented below [40–42].

- Mean Time to Repair is defined by

$$MTTR = \frac{\sum_{j=1}^{m} T_{rj} \lambda_j}{\sum_{j=1}^{m} \lambda_j} \tag{5.7}$$

where

$MTTR$ is the mean time to repair
m is the number of units.
λ_j is the constant failure rate of unit j; for $j = 1, 2, 3, \ldots, m$.
T_{rj} is the repair time required to repair unit j; for $j = 1, 2, 3, \ldots, m$.

- Maintainability Function

This measure is used to predict the probability that the repair will be completed in a time t, when it starts on an equipment/item at time $t = 0$.

Thus, the maintainability function, $M(t)$, is defined as follows:

$$M(t) = \int_0^t f(t)\,\mathrm{d}t \tag{5.8}$$

where

t is time.
$f(t)$ is the probability density function of the repair time.

Equation (5.8) is used to obtain maintainability functions for various probability distributions (*e. g.*, exponential, normal, and Weibull) representing failed equipment/item repair times. Maintainability functions for such distributions are available in Refs. [41–43].

Example 5.2

Assume that the repair times of a medical equipment are exponentially distributed with a mean value (*i. e.*, *MTTR*) of 4 hours. Thus, the probability density function of repair times is defined by

$$\begin{aligned} f(t) &= \frac{1}{MTTR} \exp\left(-\frac{t}{MTTR}\right) \\ &= \frac{1}{4} \exp\left(-\frac{t}{4}\right) \end{aligned} \tag{5.9}$$

where

$MTTR$ is the medical equipment mean time to repair.
t is time.

Calculate the probability that a repair will be completed in ten hours.

By substituting Equation (5.9) and the given data values into Equation (5.8), we get

$$M(10) = 1 - \exp\left(-\frac{10}{4}\right)$$
$$= 0.9179$$

Thus, the probability of accomplishing a repair within ten hours is 0.9179.

5.8 Organizations and Sources for Obtaining Medical Equipment Failure-related Data

There are many organizations from which failure data directly or indirectly concerned with medical equipment can be obtained. Six of these organizations are as follows:

- Center for Devices and Radiological Health (CDRH)

  Food and Drug Administration (FDA)
  1390 Piccard Drive
  Rockville, MD 20850
  USA

- Emergency Care Research Institute (ECRI)

  5200 Butler Parkway
  Plymouth Meeting, PA 19462
  USA

- Government Industry Data Exchange Program (GIDEP)

  GIDEP Operations Center
  Fleet Missile Systems, Analysis, and Evaluation Group
  Department of Navy
  Corona, CA 91720
  USA

- Parts Reliability Information Center (PRINCE)

  Reliability Office
  George C. Marshall Space Flight Center
  National Aeronautics and Space Administration (NASA)
  Huntsville, AL 35812
  USA

- Reliability Analysis Center (RAC)

  Rome Air Development Center (RADC)
  Griffiss Air Force Base
  Department of Defense
  Rome, NY 13441
  USA

- National Technical Information Service

  5285 Port Royal Road
  Springfield, VA 22161
  USA

Some of the data banks and documents for obtaining failure data concerning medical equipment are as follows:

- **Hospital Equipment Control System (HECS).** This system was developed by Emergency Care Research Institute (ECRI) in 1985 [44].
- **Medical Device Reporting System (MDRS).** This system was developed by Center for Devices and Radiological Health [45].
- **Universal Medical Device Registration and Regulatory Management System (UMDRMS).** This system was also developed by ECRI [44].
- **MIL-HDBK-217.** Reliability Prediction of Electronic Equipment, Department of Defense, Washington, D.C., USA.
- **NUREG/CR-1278.** Handbook of Human Reliability Analysis with Emphasis on Nuclear Power Plant Applications, U.S. Nuclear Regulatory Commission, Washington, D.C., USA.

5.9 Problems

1. What are the main categories of medical equipment/devices?
2. Discuss the steps of the approach developed by Bio-Optronics to produce safe and reliable medical devices.
3. Compare FMEA with FTA with respect to medical equipment.
4. List at least five facts and figures concerned, directly or indirectly, with human error in medical equipment/devices.
5. List at least 12 medical devices with a high incidence of human error.
6. Discuss important operator-related errors that occur during medical equipment/device operation or maintenance.
7. Discuss useful guidelines for reliability and other professionals to improve medical equipment reliability.
8. Define and compare the following two terms:
   - Medical equipment maintainability
   - Medical equipment maintenance

9. Discuss at least ten maintainability design factors with respect to medical equipment.
10. List at least five good sources for obtaining medical equipment reliability-related data.

References

1 Hutt, P.B., A History of Government Regulation of Adulteration and Misbranding of Medical Devices, The Medical Device Industry, Edited by N.F. Estrin, Marcel Dekker, Inc., New York, 1990, pp. 17–33.
2 Murray, K., Canada's Medical Device Industry Faces Cost Pressures, Regulatory Reform, Med. Dev. Diag. Ind. Mag., Vol. 19, No. 8, 1997, pp. 30–39.
3 Johnson, J.P., Reliability of ECG Instrumentation in a Hospital, Proceedings of the Annual Symposium on Reliability, 1967, pp. 314–318.
4 Gechman, R., Tiny Flaws in Medical Design Can Kill, Hosp. Top., Vol. 46, 1968, pp. 23–24.
5 Meyer, J.L., Some Instrument Induced Errors in the Electrocardiogram, J. Am. Med. Assoc., Vol. 201, 1967, pp. 351–358.
6 Taylor, E.F., The Effect of Medical Test Instrument Reliability on Patient Risks, Proceedings of the Annual Symposium on Reliability, 1969, pp. 328–330.
7 Crump, J.F., Safety and Reliability in Medical Electronics, Proceedings of the Annual Symposium on Reliability, 1969, pp. 320–330.
8 Dhillon, B.S., Bibliography of Literature on Medical Reliability, Microelectronics and Reliability, Vol. 20, 1980, pp. 737–742.
9 Dhillon, B.S., Reliability Engineering in Systems Design and Operation, Van Nostrand Reinhold Company, New York, 1983.
10 Dhillon, B.S., Medical Device Reliability and Associated Areas, CRC Press, Boca Raton, Florida, 2000.
11 Allen, D., California Home to Almost One-Fifth of U.S. Medical Device Industry, Med. Dev. Diag. Ind. Mag., Vol. 19, No. 10, 1997, pp. 64–67.
12 Walter, C.W., Instrumentation Failure Fatalities, Electronic News, January 27, 1969.
13 Micco, L.A., Motivation for the Biomedical Instrument Manufacturer, Proceedings of the Annual Reliability and Maintainability Symposium, 1972, pp. 242–244.
14 Banta, H.D., The Regulation of Medical Devices, Preventive Medicine, Vol. 19, 1990, pp. 693–699.
15 Medical Devices, Hearings Before the Subcommittee on Public Health and Environment, U.S. Congress House Interstate and Foreign Commerce, Serial No. 93-61, U.S. Government Printing Office, Washington, D.C., 1973.
16 Dhillon, B.S., Reliability Technology in Health Care Systems, Proceedings of the IASTED International Symposium on Computers Advanced Technology in Medicine, Health Care, and Bioengineering, 1990, pp. 84–87.
17 Kohn, L.T., Corrigan, J.M., Donaldson, M.S., Editors, To Err Is Human: Building a Safer Health System, Institute of Medicine Report, National Academy Press, Washington, D.C., 1999.

18 Schwartz, A.P., A Call for Real Added Value, Medical Industry Executive, February/March 1994, pp. 5–9.
19 Rose, H.B., A Small Instrument Manufacturer's Experience with Medical Equipment Reliability, Proceedings of the Annual Reliability and Maintainability Symposium, 1972, pp. 251–254.
20 MIL-HDBK-217, Reliability Prediction of Electronic Equipment, Department of Defense, Washington, D.C.
21 RDH-376, Reliability Design Handbook, Reliability Analysis Center, Rome Air Development Center, Griffiss Air Force Base, Rome, New York, 1976.
22 Singh, C., Reliability Calculations on Large Systems, Proceedings of the Annual Reliability and Maintainability Symposium, 1975, pp. 188–193.
23 Shooman, M.L., Probabilistic Reliability: An Engineering Approach, McGraw Hill Book Company, New York, 1968.
24 Dhillon, B.S., Design Reliability: Fundamentals and Applications, CRC Press, Boca Raton, Florida, 1999.
25 MIL-STD-1629, Procedures for Performing a Failure Mode, Effects and Criticality Analysis, Department of Defense, Washington, D.C.
26 Palady, P., Failure Modes and Effects Analysis, PT Publications, West Palm Beach, Florida, 1995.
27 Fault Tree Handbook, Report No. NUREG-0492, U.S. Nuclear Regulatory Commission, Washington, D.C., 1981.
28 Dhillon, B.S., Singh, C., Engineering Reliability: New Techniques and Applications, John Wiley and Sons, New York, 1981.
29 Bogner, M.S., Medical Devices: A New Frontier for Human Factors, CSERIAC Gateway, Vol. 4, No. 1, 1993, pp. 12–14.
30 Novel, J.L., Medical Device Failures and Adverse Effects, Pediat-Emerg. Care, Vol. 7, 1991, pp. 120–123.
31 Bogner, M.S., Medical Devices and Human Error in Human Performance in Automated Systems: Current Research and Trends, Edited by M. Mouloua and R. Parasuraman, Lawrence Erlbaum Associates, Hillsdale, New Jersey, 1994, pp. 64–67.
32 Casey, S., Set. Phasers on Stun: and Other True Tales of Design Technology and Human Error, Aegean, Inc., Santa Barbara, California, 1993.
33 Sawyer, D., Do It By Design: Introduction to Human Factors in Medical Devices, Center for Devices and Radiological Health (CDRH), Food and Drug Administration, Washington, D.C., 1996.
34 Wikland, M.E., Medical Device and Equipment Design, Interpharm press Inc., Buffalo Grove, Illinois, 1995.
35 Hyman, W.A., Human Factors in Medical Devices, in Encyclopaedia of Medical Devices and Instrumentation, edited by J.G. Webster, Vol.3, John Wiley and Sons, New York, 1988, pp. 1542–1553.
36 Taylor, E.F., The Reliability Engineer in the Health Care System, Proceedings of the Annual Reliability and Maintainability Symposium, 1972, pp. 245–248.
37 Norman, J.C., Goodman, L., Acquaintance with and Maintenance of Biomedical Instrumentation, J. Assoc. Advan. Med. Inst., Vol. 1, September 1966, pp. 8–10.
38 Waits, W., Planned Maintenance, Med. Res. Eng., Vol. 7, No. 12, 1968, pp. 15–18.
39 Cohen, T., Validating Medical Equipment Repair and Maintenance Metrics: A Progress Report, Biomedical Instrumentation and Technology, Jan./Feb., 1997, pp. 23–32.

40 Grant-Ireson, W., Coombs, C.F., Moss, R.Y., Editors, Handbook of Reliability Engineering and Management, McGraw Hill Book company, New York, 1988.
41 AMCP-133, Engineering Design Handbook: Maintainability Engineering Theory and Practice, Department of the Army, Washington, D.C., 1976.
42 Blanchard, B.S., Verma, D., Peterson, E.L., Maintainability, John Wiley and Sons, New York, 1995.
43 Dhillon, B.S., Engineering Maintainability, Gulf Publishing Company, Houston, Texas, 1999.
44 Emergency Care Research Institute (ECRI), 5200 Butler Parkway, Plymouth Meeting, Pennsylvania 19462, USA.
45 Center for Devices and Radiological Health (CDRH), Food and Drug Administration (FDA), 1390 Piccard Drive, Rockville, Maryland 20850, USA.

6

Power System Reliability

6.1 Introduction

The three main areas of an electric power system are generation, transmission, and distribution [1]. The basic function of a modern electric power system is to supply its customers cost-effective electrical energy with a high degree of reliability. During planning, design, control, operation, and maintenance of an electric power system, consideration of the two important aspects of quality and continuity of supply, along with other important factors, is normally referred to as reliability assessment. In the context of an electric power system, reliability may simply be defined as concern regarding the system's ability to provide a satisfactory amount of electrical power [2].

The history of power system reliability goes back to the early 1930s when probability concepts were applied to electric power system-related problems [3–5]. The first book on the subject in English appeared in 1970 [6]. Over the years, a large number of publications on the subject have appeared. Most of the publications on power system reliability up to 1977 are listed in Refs. [7–8]. An extensive list of recent publications on power system reliability is presented at the end of this book. This chapter presents various important aspects of power system reliability.

6.2 Terms and Definitions

There are many terms and definitions used in power system reliability. Some of the common ones are as follow: [9–12]:

- **Power system reliability.** This is the degree to which the performance of the elements in a bulk system results in electrical energy being delivered to customers within the framework of specified standards and in the amount required.
- **Forced outage.** This is when a piece of equipment or a unit has to be taken out of service because of damage or a component failure.

- **Forced derating.** This is when a piece of equipment or a unit is operated at a forced derated or lowered capacity because of damage or a component failure.
- **Scheduled outage.** This is the shutdown of a generating unit, transmission line, or other facility, for maintenance or inspection, as per an advance schedule.
- **Service hours.** These are the total number of operation hours of a piece of equipment or a unit.
- **Forced outage hours.** These are the total number of hours a piece of equipment or a unit spends in the forced outage condition.
- **Mean time to forced outage.** This is analogous to mean time to failure (MTTF) and is given by the total of service hours over the total number of forced outages.
- **Mean forced outage duration.** This is analogous to mean time to repair (MTTR) and is given by the total number of forced outage hours over the total number of forced outages.
- **Forced outage rate.** This is (for an equipment) given by the total number of forced outage hours times 100 over the total number of service hours plus the total number of forced outage hours.

6.3 Service Performance Indices

In the electric power system area, usually various service performance indices are calculated for the total system, a specific region or voltage level, designated feeders or different groups of customers, etc. [2]. Some of the most widely used indices are presented below [2, 13].

6.3.1 Index I

The Index I is known as average service availability index (ASAI) and is expressed by

$$ASAI = \frac{CHAS}{CHD} \tag{6.1}$$

where

$CHAS$ is the customer hours of available service.

CHD is customer hours demanded. These hours are given by the 12-month average number of customers serviced times 8,760 hours.

6.3.2 Index II

The Index II is known as system average interruption frequency index (SAIFI) and is defined by

$$SAIFI = \frac{TNCI}{TNC} \tag{6.2}$$

where
TNC is the total number of customers.
$TNCI$ is the total number of customer interruptions per year.

6.3.3 Index III

The Index III is known as system average interruption duration index (SAIDI) and is expressed by

$$SAIDI = \frac{SCID}{TNC} \tag{6.3}$$

where
$SCID$ is the sum of customer interruption durations per year.

6.3.4 Index IV

The Index IV is known as customer average interruption frequency index (CAIFI) and is defined by

$$CAIFI = \frac{TNCI}{TNCA} \tag{6.4}$$

where
$TNCA$ is the total number of customers affected. It is be noted that the customers affected should only be counted once, irrespective of the number of interruptions during the year they may have experienced.

6.3.5 Index V

The Index V is known as customer average interruption duration index (CAIDI) and is expressed by

$$\begin{aligned} CAIDI &= \frac{SAIDI}{DAIFI} \\ &= \frac{SCID}{TNCI} \end{aligned} \tag{6.5}$$

6.4 Loss of Load Probability

Over the years, loss-of-load probability (LOLP) has been used as the single most important metric for estimating overall power system reliability. LOLP may simply be described as a projected value of how much time, in the long run, the load on a given power system is expected to be more than the capacity of the generating resources [9]. Various probabilistic techniques are used to calculate LOLP.

In the setting up of an LOLP criterion, it is assumed that an electric power system strong enough to have a low LOLP can probably withstand most of the foreseeable peak loads, outages, and contingencies. Thus, an utility is expected to arrange for resources (*i. e.*, generation, purchases, load management, and so on) in such a way so that the resulting system LOLP will be at or less than an acceptable level.

Usually, the common practice is to plan to power system for achieving an LOLP of 0.1 days per year or less. All in all, some of the difficulties with this use of LOLP are as follows [9]:

- Different LOLP estimation methods can lead to different indices for exactly the same electric power system.
- LOLP itself does not specify the magnitude or duration of the shortage of electricity.
- Major loss-of-load incidents normally occur because of contingencies not modeled properly by the traditional LOLP calculation.
- LOLP does not take into consideration the factor of additional emergency support that one region or control area may receive from another, or other emergency actions/measures that control area operators can exercise to maintain system reliability.

6.5 Models for Performing Availability Analysis of a Single Generator Unit

There are a number of mathematical models that can be used to perform availability analysis of a single generator unit. This section presents three Markov models that can also be used to perform availability analysis of equipment other than a generator unit [12]. Two examples of such equipment are a transformer and a pulverizer.

6.5.1 Model I

Model I represents a generator unit that can either be in operating state or in failed state. The failed generator unit is repaired. The generator unit state space diagram is shown in Figure 6.1. The numerals in the rectangle and circle denote the system state. The following assumptions are associated with the model:

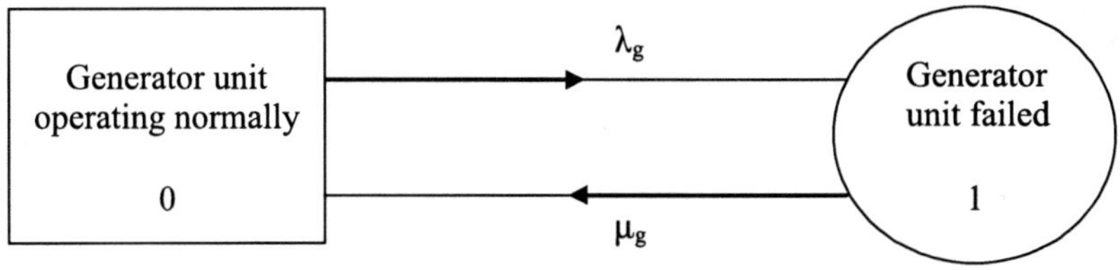

Figure 6.1. Generator unit state space diagram

- The generator unit failures are statistically independent.
- The generator unit failure and repair rates are constant.
- The repaired generator unit is as good as new.

The following symbols are associated with the diagram in Figure 6.1 and its associated equations:

$P_i(t)$ is the probability that the generator unit is in state i at time t; for $i=0$ (operating normally), $i=1$ (failed).
λ_g is the generator unit failure rate.
μ_g is the generator unit repair rate.

Using the Markov method, we write down the following equations for the diagram in Figure 6.1 diagram [1, 12]:

$$\frac{dP_0(t)}{dt} + \lambda_g P_0(t) - \mu_g P_1(t) = 0 \tag{6.6}$$

$$\frac{dP_1(t)}{dt} + \mu_g P_1(t) - \lambda_g P_0(t) = 0 \tag{6.7}$$

At time $t=0$, $P_0(0)=1$ and $P_1(0)=0$

Solving Equations (6.6)–(6.7) by using Laplace transforms we get

$$P_0(t) = \frac{\mu_g}{\lambda_g + \mu_g} + \frac{\lambda_g}{\lambda_g + \mu_g} e^{-(\lambda_g + \mu_g)t} \tag{6.8}$$

$$P_1(t) = \frac{\lambda_g}{\lambda_g + \mu_g} - \frac{\mu_g}{\lambda_g + \mu_g} e^{-(\lambda_g + \mu_g)t} \tag{6.9}$$

The generator unit availability and unavailability are given by

$$AV_g(t) = P_0(t) = \frac{\mu_g}{\lambda_g + \mu_g} + \frac{\lambda_g}{\lambda_g + \mu_g} e^{-(\lambda_g + \mu_g)t} \tag{6.10}$$

and

$$UA_g(t) = P_1(t) = \frac{\lambda_g}{\lambda_g + \mu_g} - \frac{\mu_g}{\lambda_g + \mu_g} e^{-(\lambda_g + \mu_g)t} \tag{6.11}$$

where
$AV_g(t)$ is the generator unit availability at time t.
$UA_g(t)$ is the generator unit unavailability at time t.

For large t, Equations (6.10)–(6.11) reduce to

$$AV_g = \frac{\mu_g}{\lambda_g + \mu_g} \tag{6.12}$$

and

$$UA_g = \frac{\lambda_g}{\lambda_g + \mu_g} \tag{6.13}$$

where
AV_g is the generator unit steady state availability.
UA_g is the generator unit steady state unavailability.

Since $\lambda_g = \frac{1}{MTTF_g}$ and $\mu_g = \frac{1}{MTTR_g}$, Equations (6.12)–(6.13) become

$$AV_g = \frac{MTTF_g}{MTTR_g + MTTF_g} = \frac{\text{Generator unit uptime}}{\text{Generator unit downtime} + \text{Generator unit uptime}} \tag{6.14}$$

and

$$UA_g = \frac{MTTR_g}{MTTR_g + MTTF_g} = \frac{\text{Generator unit downtime}}{\text{Generator unit downtime} + \text{Generator unit uptime}} \tag{6.15}$$

where
$MTTF_g$ is the generator unit mean time to failure.
$MTTR_g$ is the generator unit mean time to repair.

Example 6.1
Assume that a generator unit's constant failure and repair rates are as follows:

$$\lambda_g = 0.0004 \text{ failures / hour}$$

and

$$\mu_g = 0.0009 \text{ repairs / hour}$$

Calculate the generator unit's steady state availability. By substituting the given data values into Equation (6.12) we get

$$AV_g = \frac{0.0009}{0.0004 + 0.0009} = 0.6923$$

Thus, the generator unit's steady state availability is 0.6923.

6.5.2 Model II

Model II represents a generator unit that can either be in operating state or failed state or down for preventive maintenance. This is depicted by the state space diagram shown in Figure 6.2. The numerals in the rectangle, diamond, and circle denote the system state.

The following assumptions are associated with the model:

- The generator unit failures are statistically independent.
- The generator unit failure, repair, preventive maintenance down, and preventive maintenance performance rates are constant.
- After repair and preventive maintenance the generator unit is as good as new.

The following symbols are associated with the diagram in Figure 6.2 and its associated equations:

$P_i(t)$ is the probability that the generator unit is in state i at time t; for $i=0$ (operating normally), $i=1$ (down for preventive maintenance), $i=2$ (failed).
λ is the generator unit failure rate.
μ is the generator unit repair rate.
λ_p is the generator unit (down for) preventive maintenance rate.
μ_p is the generator unit preventive maintenance performance (repair) rate.

As for Model I, using the Markov method, we write down the following equations for the Figure 6.2 diagram [1, 12]:

$$\frac{dP_0(t)}{dt}+(\lambda_p+\lambda)P_0(t)-\mu P_2(t)-\lambda_p P_1(t)=0 \tag{6.16}$$

$$\frac{dP_1(t)}{dt}+\mu_p P_1(t)-\lambda_p P_0(t)=0 \tag{6.17}$$

$$\frac{dP_2(t)}{dt}+\mu P_2(t)-\lambda P_0(t)=0 \tag{6.18}$$

At time $t=0$, $P_0(0)=1$, $P_1(0)=0$, and $P_2(0)=0$.

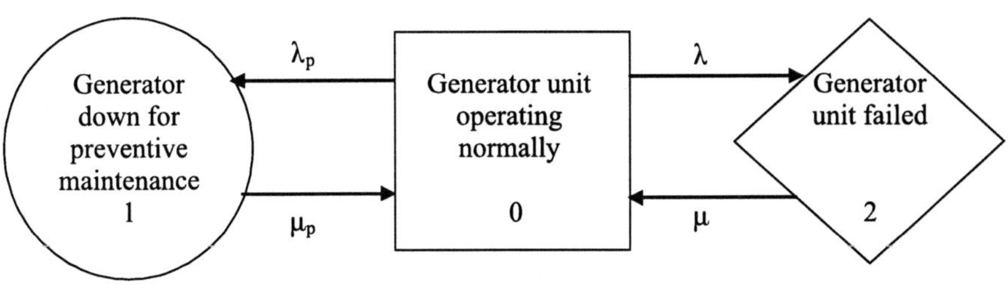

Figure 6.2. Generator unit state space diagram

Solving Equations (6.16)–(6.18) by using Laplace transforms, we get

$$P_0(t) = \frac{\mu_p \mu}{c_1 c_2} + \left[\frac{(c_1 + \mu_p)(c_1 + \mu)}{c_1 (c_1 - c_2)}\right] e^{c_1 t} - \left[\frac{(c_2 + \mu_p)(c_2 + \mu)}{c_2 (c_1 - c_2)}\right] e^{c_2 t} \tag{6.19}$$

$$P_1(t) = \frac{\lambda_p \mu}{c_1 c_2} + \left[\frac{(\lambda_p c_1 + \lambda_p \mu)}{c_1 (c_1 - c_2)}\right] e^{c_1 t} - \left[\frac{(\mu + c_2)\lambda_p}{c_2 (c_1 - c_2)}\right] e^{c_2 t} \tag{6.20}$$

$$P_2(t) = \frac{\lambda \mu_p}{c_1 c_2} + \left[\frac{(\lambda c_1 + \lambda \mu_p)}{c_1 (c_1 - c_2)}\right] e^{c_1 t} - \left[\frac{(\mu_p + c_2)\lambda}{c_2 (c_1 - c_2)}\right] e^{c_2 t} \tag{6.21}$$

where

$$c_1 c_2 = \mu_p \mu + \lambda_p \mu + \lambda \mu_p \tag{6.22}$$

$$c_1 + c_2 = -(\mu_p + \mu + + \lambda_p + \lambda) \tag{6.23}$$

The generator unit availability, $AV_g(t)$, is given by

$$AV_g(t) = P_0(t) = \frac{\mu_p \mu}{c_1 c_2} + \left[\frac{(c_1 + \mu_p)(c_1 + \mu)}{c_1 (c_1 - c_2)}\right] e^{c_1 t} - \left[\frac{(c_2 + \mu_p)(c_2 + \mu)}{c_2 (c_1 - c_2)}\right] e^{c_2 t} \tag{6.24}$$

The above availability expression is valid if and only if c_1 and c_2 are negative. Thus, for large t, Equation (6.24) reduces to

$$AV_g = \lim_{t \to \infty} AV_g(t) = \frac{\mu_p \mu}{c_1 c_2} \tag{6.25}$$

where
AV_g is the generator unit steady state availability.

Example 6.2
Assume that for a generator unit we have the following data values:

$\lambda = 0.0002$ failures / hour
$\lambda_p = 0.0005$ / hour
$\mu = 0.0006$ repairs / hour
and $\mu_p = 0.0009$ / hour

Calculate the generator unit's steady state availability. By substituting the specified data values into Equation (6.25), we get

$$AV_g = \frac{(0.0009)(0.0006)}{(0.0009)(0.0006) + (0.0005)(0.0006) + (0.0002)(0.0009)}$$
$$= 0.5294$$

Thus, the generator unit's steady state availability is 0.5294.

6.5.3 Model III

Model III represents a power generator unit that can be either operating normally (*i. e.*, producing electricity at its full capacity), derated (*i. e.*, producing electricity at a derated capacity, for example, say 250 megawatts instead of 500 megawatts at full capacity), or failed. This is depicted by the state space diagram in Figure 6.3. The numerals in the rectangle, circle, and diagram denote system state.

The following assumptions are associated with the model:

- The generator unit failures are statistically independent.
- The repaired generator unit is as good as new.
- All generator unit failure and repair rates are constant.

The following symbols are associated with the diagram in Figure 6.3 and its associated equations:

$P_i(t)$ is the probability that the generator unit is in state i at time t; for $i=0$ (operating normally), $i=1$ (derated), $i=2$ (failed).
λ is the generator unit failure rate from state 0 to state 2.
λ_d is the generator unit failure rate from state 0 to state 1.
λ_1 is the generator unit failure rate from state 1 to state 2.
μ is the generator unit repair rate from state 2 to state 0.
μ_d is the generator unit repair rate from state 1 to state 0.
μ_1 is the generator unit repair rate from state 2 to state 1.

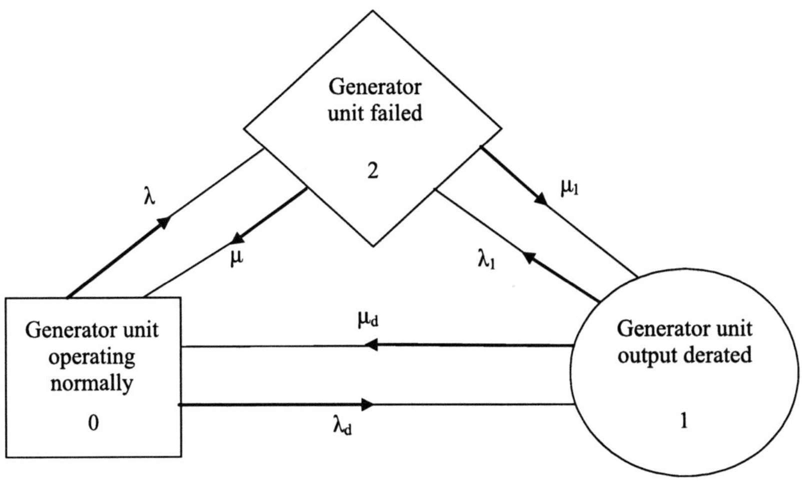

Figure 6.3. Generator unit state space diagram

As for Models I and II, using the Markov method, we write down the following equations for Figure 6.3 diagram [1, 12]:

$$\frac{dP_0(t)}{dt} + (\lambda_d + \lambda)\,P_0(t) - \mu_d\,P_1(t) - \mu\,P_2(t) = 0 \tag{6.26}$$

$$\frac{dP_1(t)}{dt} + (\mu_d + \lambda_1)\,P_1(t) - \mu_1\,P_2(t) - \lambda_d\,P_0(t) = 0 \tag{6.27}$$

$$\frac{dP_2(t)}{dt} + (\mu + \mu_1)\,P_2(t) - \lambda_1\,P_1(t) - \lambda\,P_0(t) = 0 \tag{6.28}$$

At time $t = 0$, $P_0(0) = 1$, $P_1(0) = 0$, and $P_2(0) = 0$.

Solving Equations (6.26)–(6.28) by using Laplace transforms, we get

$$P_0(t) = \frac{A_1}{k_1 k_2} + \frac{A_2}{k_1(k_1 - k_2)}\,e^{k_1 t} + \left[1 - \frac{A_1}{k_1 k_2} - \frac{A_2}{k_1(k_1 - k_2)}\right] e^{k_2 t} \tag{6.29}$$

where

$$A_1 = \mu\mu_d + \lambda_1\mu + \mu_d\,\mu_1 \tag{6.30}$$

$$A_2 = \mu_d\,k_1 + \mu k_1 + \mu_1\,k_1 + k_1\lambda_1 + k_1^2 + \mu_d\,\mu + \lambda_1\mu + \mu_d\,\mu_1 \tag{6.31}$$

$$k_1, k_2 = \frac{-A_3 \pm \left[A_3^2 - 4A_4\right]^{1/2}}{2} \tag{6.32}$$

$$A_3 = \mu + \mu_1 + \mu_d + \lambda + \lambda_1 + \lambda_d \tag{6.33}$$

$$A_4 = \mu_d\,\mu + \lambda_1\mu + \mu_d\,\mu_1 + \mu\lambda_d + \lambda_1\lambda_d + \mu_d\lambda + \lambda\mu_1 + \lambda\lambda_1 + \lambda_d\,\mu_1 \tag{6.34}$$

$$P_1(t) = \frac{A_5}{k_1 k_2} + \frac{A_6}{k_1(k_1 - k_2)}\,e^{k_1 t} - \left[\frac{A_5}{k_1 k_2} + \frac{A_6}{k_1(k_1 - k_2)}\right] e^{k_2 t} \tag{6.35}$$

where

$$A_5 = \lambda_d\,\mu + \lambda_d\,\mu_1 + \lambda\mu_1 \tag{6.36}$$

$$A_6 = k_1\lambda_d + A_5 \tag{6.37}$$

$$P_2(t) = \frac{A_7}{k_1 k_2} + \frac{A_8}{k_1(k_1 - k_2)}\,e^{k_1 t} - \left[\frac{A_7}{k_1 k_2} + \frac{A_8}{k_1(k_1 - k_2)}\right] e^{k_2 t} \tag{6.38}$$

where

$$A_7 = \lambda_d\lambda_1 + \mu_d\lambda + \lambda\lambda_1 \tag{6.39}$$

$$A_8 = k_1\lambda + A_7 \tag{6.40}$$

The generator unit operational availability is given by

$$AV_{go}(t) = P_0(t) + P_1(t) \tag{6.41}$$

For large t, Equation (6.41) reduces to

$$AV_{go} = \lim_{t \to \infty} \left[P_0(t) + P_1(t)\right] = \frac{A_1 + A_5}{k_1 k_2} \tag{6.42}$$

where

AV_{go} is the generator unit operational steady state availability.

6.6 Models for Performing Availability Analysis of Transmission and Associated Systems

In the power system area various types of equipment and systems are used to transmit electrical energy from one end to another. Two examples of such systems are transmission lines and transformers. This section presents three Markov models for performing availability analysis of transmission lines and transformers [6, 11, 12].

6.6.1 Model I

Model I represents transmission lines and other equipment operating in fluctuating outdoor environments (*i. e.*, normal and stormy). The system can fail under both these conditions. The system state space diagram is shown in Figure 6.4. The numerals in rectangles and circles denote system states. The following assumptions are associated with the model:

- All failures are statistically independent.
- All failures, repair, and weather fluctuation rates are constant.
- The repaired system is as good as new.

The following symbols are associated with the diagram in Figure 6.4 and its associated equations:

$P_i(t)$ is the probability that the system is in state i at time t; for $i=0$ (operating normally in normal weather), $i=1$ (failed in normal weather), $i=2$ (operating normally in stormy weather), $i=3$ (failed in stormy weather).
α is the constant transition rate from normal weather to stormy weather.
β is the constant transition rate from stormy weather to normal weather.
λ_n is the system constant failure rate in normal weather.
λ_s is the system constant failure rate in stormy weather.
μ_n is the system constant failure rate in stormy weather.
μ_s is the system constant repair rate in stormy weather.

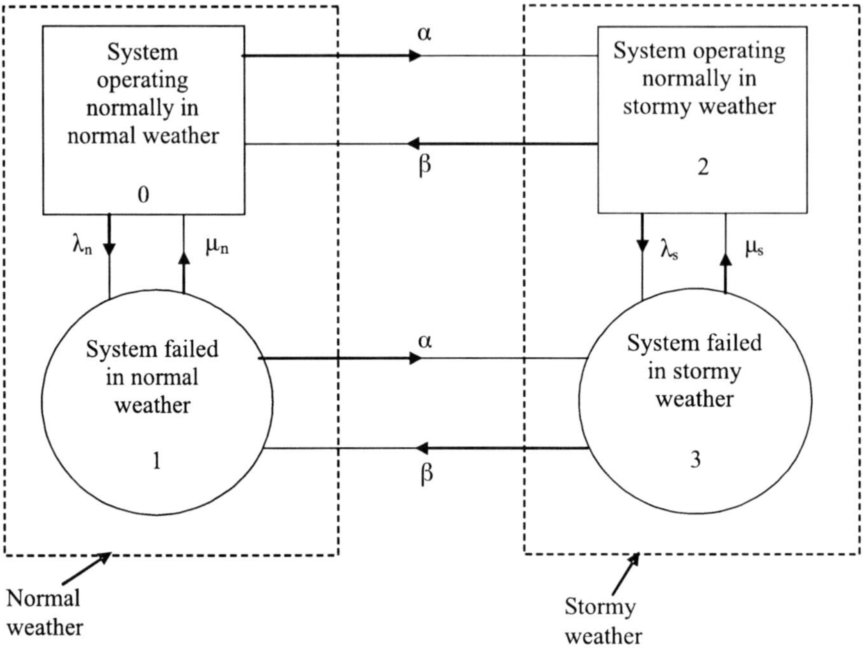

Figure 6.4. State space diagram of a system operating under fluctuating environments

Using the Markov method, we write down the following equations for Figure 6.4 diagram [1, 12]:

$$\frac{dP_0(t)}{dt} + (\lambda_n + \alpha) P_0(t) - \beta P_2(t) - \mu_n P_1(t) = 0 \tag{6.43}$$

$$\frac{dP_1(t)}{dt} + (\mu_n + \alpha) P_1(t) - \beta P_3(t) - \lambda_n P_0(t) = 0 \tag{6.44}$$

$$\frac{dP_2(t)}{dt} + (\lambda_s + \beta) P_2(t) - \mu_s P_3(t) - \alpha P_0(t) = 0 \tag{6.45}$$

$$\frac{dP_3(t)}{dt} + (\beta + \mu_s) P_3(t) - \lambda_s P_2(t) - \alpha P_1(t) = 0 \tag{6.46}$$

At time $t = 0$, $P_0(0) = 1$, $P_1(0) = 0$, $P_2(0) = (0)$, and $P_3(0) = 0$.

The following steady-state equations are obtained from Equations (6.43)–(6.46) by setting the derivatives with respect to time t equal to zero and using the relationship $\sum_{i=0}^{3} P_i = 1$:

$$P_o = \frac{\beta B_1}{\alpha\left(B_2 + B_3\right) + \beta\left(B_4 + B_1\right)} \tag{6.47}$$

where

$$B_1 = \mu_s \alpha + \mu_n B_5 \tag{6.48}$$

$$B_2 = \mu_n \beta + \mu_s B_6 \tag{6.49}$$

$$B_3 = \lambda_n \beta + \lambda_s B_6 \tag{6.50}$$

$$B_4 = \lambda_s \alpha + \lambda_n B_5 \tag{6.51}$$

$$B_5 = \lambda_s + \beta + \mu_s \tag{6.52}$$

$$B_6 = \lambda_n + \alpha + \mu_n \tag{6.53}$$

$$P_1 = B_4 P_0 / B_1 \tag{6.54}$$

$$P_2 = \alpha P_0 B_2 / \beta B_1 \tag{6.55}$$

$$P_3 = \alpha P_0 B_3 / \beta B_1 \tag{6.56}$$

P_0, P_1, P_2, and P_3 are the steady state probabilities of the system being in states 0, 1, 2, and 3, respectively.

The system steady state availability, AV_{ss}, is given by

$$AV_{ss} = P_0 + P_2 \tag{6.57}$$

6.6.2 Model II

Model II represents a system composed of two non-identical and redundant transmission lines subject to common-cause failures. A common-cause failure may simply be described as any instance where multiple units fail due to a single cause [14, 15]. In transmission lines a common-cause failure may occur due to factors such as poor weather, tornado, and aircraft crash. The system state space diagram is shown in Figure 6.5. The numerals in the boxes and the circle denote system states.

The following assumptions are associated with the model:

- All failures are statistically independent.
- A repaired transmission line is as good as new.
- All failures and repair rates are constant.

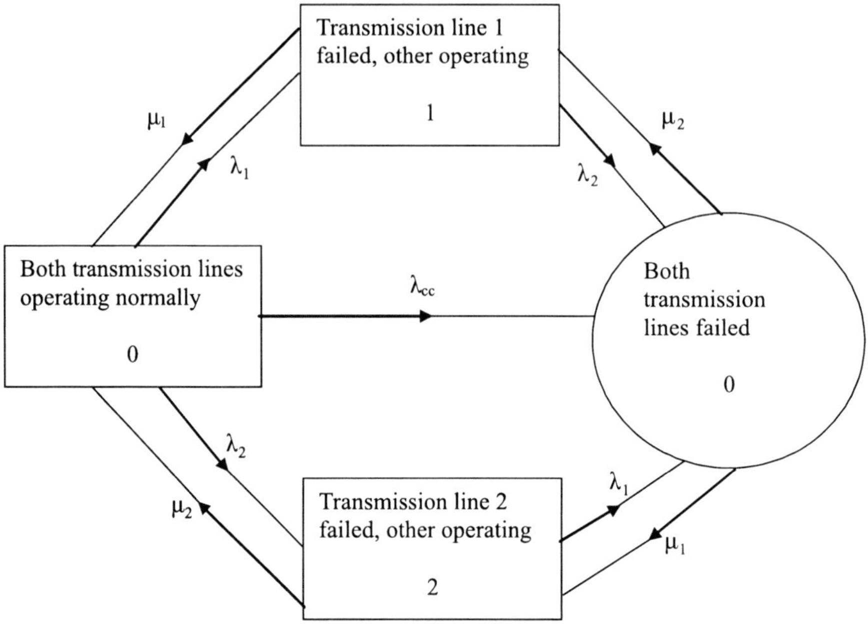

Figure 6.5. State space diagram for two non-identical transmission lines

The following symbols are associated with Figure 6.5 diagram and its associated equations:

$P_i(t)$ is the probability that the system is in state i at time t; for $i=0$ (both transmission lines operating normally), $i=1$ (transmission line 1 failed, other operating), $i=2$ (transmission line 2 failed, other operating), $i=3$ (both transmission lines failed).

λ_1 is the transmission line 1 failure rate.

λ_2 is the transmission line 2 failure rate.

λ_{cc} is the system common-cause failure rate.

μ_1 is the transmission line 1 repair rate.

μ_2 is the transmission line 2 repair rate.

As for Model I, using the Markov method, we write down the following equations for Figure 6.5 diagram [1, 12, 15]:

$$\frac{dP_0(t)}{dt}+(\lambda_1+\lambda_2+\lambda_{cc})P_0(t)-\mu_1 P_1(t)-\mu_2 P_2(t)=0 \tag{6.58}$$

$$\frac{dP_1(t)}{dt}+(\lambda_2+\mu_1)P_1(t)-\mu_2 P_3(t)-\lambda_1 P_0(t)=0 \tag{6.59}$$

$$\frac{dP_2(t)}{dt}+(\lambda_1+\mu_2)P_2(t)-\mu_1 P_3(t)-\lambda_2 P_0(t)=0 \tag{6.60}$$

$$\frac{dP_3(t)}{dt} + (\mu_1 + \mu_2) P_3(t) - \lambda_1 P_2(t) - \lambda_2 P_1(t) - \lambda_{cc} P_0(t) = 0 \tag{6.61}$$

At time $t = 0$, $P_0(0) = 1$, $P_1(0) = 0$, $P_2(0) = 0$, and $P_3(0) = 0$.

The following steady-state equations are obtained from Equations (6.58)–(6.61) by setting the derivatives with respect to time t equal to zero and using the relationship $\sum_{i=0}^{3} P_i = 1$:

$$P_o = \mu_1 \mu_2 C / C_3 \tag{6.62}$$

where

$$C = C_1 + C_2 \tag{6.63}$$

$$C_1 = (\lambda_1 + \mu_1) \tag{6.64}$$

$$C_2 = (\lambda_2 + \mu_2) \tag{6.65}$$

$$C_3 = CC_1 C_2 + \lambda_{CC} \left[C_1 (C_2 + \mu_1) + \mu_2 C_2 \right] \tag{6.66}$$

$$P_1 = [C_1 \lambda_1 + C_4 \lambda_{cc}] \mu_2 / C_3 \tag{6.67}$$

where

$$C_4 = (\lambda_1 + \mu_2) \tag{6.68}$$

$$P_2 = [C \lambda_2 + C_5 \lambda_{cc}] \mu_1 / C_3 \tag{6.69}$$

where

$$C_5 = (\lambda_2 + \mu_1) \tag{6.70}$$

$$P_3 = C \lambda_1 \lambda_2 + C_4 C_5 \lambda_{cc} / C_3 \tag{6.71}$$

P_0, P_1, P_2, and P_3 are the steady state probabilities of the system being in states 0, 1, 2, and 3, respectively.

The system steady state availability, AV_{ss}, is given by

$$AV_{ss} = P_0 + P_1 + P_2 \tag{6.72}$$

6.6.3 Model III

Model III represents a system composed of three active and identical single-phase transformers with one standby transformer (*i. e.*, unit) [11]. The system state space diagram is shown in Figure 6.6. The numerals in boxes denote system states.

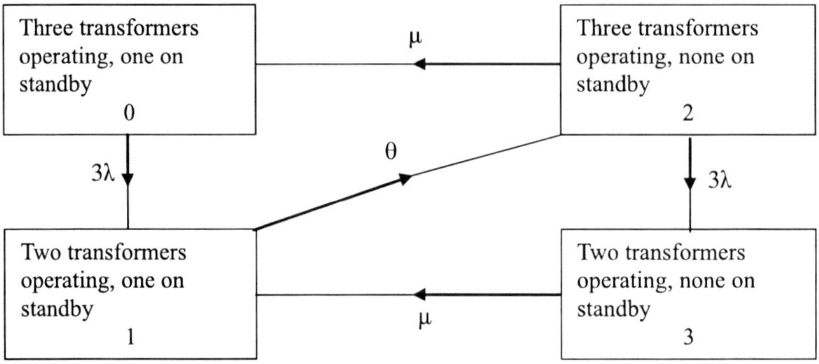

Figure 6.6. State space diagram of three single-phase transformers with one standby

The model is subject to the following assumptions [11, 12]:

- Transformer failure, repair, and replacement (*i. e.*, installation) rates are constant.
- All failures are statistically independent.
- The standby transformer or unit cannot fail in its standby mode.
- The whole transformer bank is considered failed when more than one transformer fails. In addition, it is assumed that no more transformer failures occur.
- Repaired transformers are as good as new.

The following symbols are associated with the diagram in Figure 6.6 and its associated equations:

$P_i(t)$ is the probability that the system is in state i at time t; for $i=0$ (three transformers operating, one on standby), $i=1$ (two transformers operating, one on standby), $i=2$ (three transformers operating, none on standby), $i=3$ (two transformers operating, none on standby).

λ is the transformer failure rate.

μ is the transformer repair rate.

θ is the standby transformer or unit installation rate.

As for Models I and II, using the Markov method, we write down the following equations for Figure 6.6 diagram [1, 11, 12]:

$$\frac{dP_0(t)}{dt} + 3\lambda P_0(t) - \mu P_2(t) = 0 \tag{6.73}$$

$$\frac{dP_1(t)}{dt} + \theta P_1(t) - 3\lambda P_0(t) - \mu P_3(t) = 0 \tag{6.74}$$

$$\frac{dP_2(t)}{dt} + (3\lambda + \mu) P_2(t) - \theta P_1(t) = 0 \tag{6.75}$$

$$\frac{dP_3(t)}{dt} + \mu P_3(t) - 3\lambda P_2(t) = 0 \tag{6.76}$$

At time $t = 0$, $P_0(0) = 1$, $P_1(0) = 0$, $P_2(0) = 0$, and $P_3(0) = 0$.

The following steady-state equations are obtained from Equations (6.73)–(6.76) by setting the derivatives with respect to time t equal to zero and using the relationship $\sum_{i=0}^{3} P_i = 1$:

$$P_0 = \left[1 + D_1\left(1 + D_2 + D_1\right)\right]^{-1} \tag{6.77}$$

where

$$D_1 = 3\lambda / \mu \tag{6.78}$$

$$D_2 = (3\lambda + \mu)/\theta \tag{6.79}$$

$$P_1 = D_1 D_2 P_0 \tag{6.80}$$

$$P_2 = D_1 P_0 \tag{6.81}$$

$$P_3 = D_1^2 P_0 \tag{6.82}$$

P_0, P_1, P_2, and P_3 are the steady state probabilities of the system being in states 0, 1, 2, and 3, respectively.

6.7 Problems

1. Write an essay on power system reliability.
2. Define the following terms:
   - Power system reliability
   - Forced outage rate
   - Forced derating
3. Define the following indices:
   - SAIFI
   - CAIDI
   - CAIFI
4. What is loss of load probability (LOLP) ?
5. What are the problems associated with the use of LOLP?
6. A generator unit's constant failure and repair rates are as follows:
   - $\lambda = 0.003$ failure/hour
   - $\mu = 0.008$ repairs/hour

   Calculate the generator unit's steady state availability.

7. Using the data from problem 6, calculate the generator unit's steady state unavailability.
8. Prove Equation (6.42).
9. Prove that the sum of Equations (6.47), (6.54)–(6.56) is equal to unity.
10. Prove Equations (6.62), (6.67), (6.69), and (6.71).

References

1 Dhillon, B.S., Singh, C., Engineering Reliability: New Techniques and Applications, John Wiley and Sons, New York, 1981.
2 Billinton, R., Allan, R.N., Reliability of Electric Power Systems: An Overview, in Handbook of Reliability Engineering, edited by H. Pham, Springer-Verlag, London, 2003, pp. 511–528.
3 Layman, W.J., Fundamental Consideration in Preparing a Master System Plan, Electrical World, Vol. 101, 1933, 778–792.
4 Smith, S.A., Service Reliability Measured by Probabilities of Outage, Electrical World, Vol. 103, 1934, pp. 371–374.
5 Dhillon, B.S., Power System Reliability, Safety, and Management, Ann Arbor Science Publishers, Ann Arbor, Michigan, 1983.
6 Billinton, R., Power System Reliability Evaluation, Gordon and Breach Science Publishers, New York, 1970.
7 Billinton, R., Bibliography on the Application of Probability Methods in Power System Reliability Evaluation, IEEE Transactions on Power Apparatus and Systems, Vol. 91, 1972, pp. 649–660.
8 "Bibliography on the Application of Probability Methods in Power System Reliability Evaluation", IEEE Transactions on Power Apparatus and Systems, Vol. 97, 1978, pp. 2235–2242.
9 Kueck, J.D., Kirby, B.J., Overholt, P.N., Markel, L.C., Measurement Practices for Reliability and Power Quality, Report No. ORNL/TM-2004/91, June 2004. Available from the Oak Ridge National Laboratory, Oak Ridge, Tennessee, USA.
10 Kennedy, B., Power Quality Primer, McGraw Hill Book Company, New York, 2000.
11 Endrenyi, J., Reliability Modeling in Electric Power Systems, John Wiley and Sons, New York, 1978.
12 Dhillon, B.S., Reliability Engineering in Systems Design and Operation, Van Nostrand Reinhold Company, New York, 1983.
13 Billinton, R., Allan, R.N., Reliability Evaluation of Power Systems, Plenum Press, New York, 1996.
14 Gangloff, W.C., Common Mode Failure Analysis, IEEE Transactions on Power Apparatus and Systems, Vol. 94, Feb. 1975, pp. 27–30.
15 Billinton, R., Medicherla, T.L.P., Sachdev, M.S., Common-Cause Outages in Multiple Circuit Transmission Lines, IEEE Transactions on Reliability, Vol. 27, 1978, pp. 128–131.

7

Computer and Internet Reliability

7.1 Introduction

Today, billions of dollars are being spent annually to produce computers for various types of applications ranging from personal use to control space and other systems. As the computers are composed of both the hardware and software components, the reliability of both these components is equally important for their successful operation. The history of computer hardware reliability may be traced back to the works of Shannon [1], Hamming [2], Von Neumann [3] and Moore and Shannon [4] that appeared in 1948, 1950, 1956, and 1956, respectively. For example, in 1956 Von Neumann proposed the well-known triple modular redundancy (TMR) scheme to improve hardware reliability.

It appears that the first serious effort on software reliability started at Bell Laboratories in 1964 [5]. Nonetheless, some of the important works that appeared in the 1960s were by Haugk, Tsiang, and Zimmerman [6], Floyd [7], Hudson [8], Barlow and Scheuer [9], London [10], and Sauter [11]. Computer hardware and software reliability history is discussed in detail in Ref. [12].

The history of the Internet goes back to 1969 with the development of Advanced Research Projects Agency Network (ARPANET). It has grown from 4 hosts in 1969 to over 147 million hosts and 38 million sites in 2002. In 2000, the Internet economy generated around $830 billion in revenues in the United States alone. In 2001, there were 52,658 Internet-related incidents and failures. Needless to say, today a reliable and stable Internet is extremely important to the global economy and other areas, because Internet failures can easily generate millions of dollars in losses and interrupt the daily routine of hundreds of thousands of end users [13]. An extensive list of references directly or indirectly related to Internet reliability are listed at the end of this book. This chapter presents various important aspects of computer hardware, software, and Internet reliability.

7.2 Computer System Failure Causes and Reliability Measures

Although there are many causes for computer system failures, the important ones are as follows [12, 14, 15]:

- Human errors
- Processor and memory failures
- Peripheral device failures
- Environmental and power failures
- Communication network failures
- Saturation
- Gradual erosion of the data base
- Mysterious failures

Some of the above causes or sources of computer system failure are described below.

Human errors, in general, occur due to operator mistakes and oversights, and often occur during starting up, running, and shutting down the system. Processor and memory failures are associated with processor and memory party errors. Although processor errors occur quite rarely, they are generally catastrophic. However, there are occasions when the central processor fails to execute instructions properly due to a "dropped bit". Nowadays, the memory parity errors occur very rarely because of improvements in hardware reliability and also they are not necessarily fatal.

Peripheral device failures are quite important to consider because they too can cause serious problems but they seldom lead to a system shutdown. The frequently occurring errors in peripheral devices are transient or intermittent, and the electromechanical nature of the devices is the usual reason for their occurrence.

Environmental failures occur due to factors such as failure of air conditioning equipment, fires, earthquakes, and electromagnetic interference, whereas power failures due to factors such as transient fluctuations in frequency or voltage and total power loss from the local utility company. Communication network failures are mostly of a transient nature and are associated with inter-module communication. The use of "vertical parity" logic can help to cut down around 70% of errors in communication lines.

In real-life systems, the failures that cannot be categorized properly are known as mysterious failures. An example of such failures is the sudden stop functioning of a normally operating system without indication of any problem (*i. e.*, software, hardware, etc.).

There are many measures used in performing computer system reliability analysis. They may be grouped under two broad categories: Category I and Category II. Category I includes measures such as system reliability, system availability, mean time to failure, and mission time. These measures are suitable for configurations such as standby, hybrid, and massive redundant systems [3, 12, 16–18]. Category II includes measures such as computation reliability (*i. e.*, the failure-free probability that the system will without an error execute a task of length x initiated at time t), computation availability (*i. e.*, the expected computation capacity of the system at given time t), mean computation before failure (*i. e.*, the expected amount of computation available on the system before failure), capacity threshold (*i. e.*, that time at which a certain value of computation availability is reached), and computation threshold (*i. e.*, the time at which a certain value of computation reliability is reached for a task whose length is x) to handle gracefully degrading systems [12, 19].

7.3 Comparisons Between Hardware and Software Reliability

As it is important to have a clear understanding of the differences between computer hardware and software reliability, Table 7.1 presents comparisons of the some important areas [20–22].

Table 7.1. Hardware and software reliability comparisons

Hardware reliability	Software reliability
Wears out.	Does not wear out.
A hardware failure is usually caused by physical effects.	A software failure is caused by programming error.
Normally redundancy is quite effective.	Redundancy may not be effective at all.
Failure of many hardware components is governed by the "bathtub" hazard rate curve.	Software failures are not governed by the "bathtub" hazard rate curve.
Obtaining good quality failure data is a problem.	Obtaining good quality failure data is a problem.
Interfaces are visual.	Interfaces are conceptual.
The failed system is repaired back to its operating state by performing corrective maintenance.	Corrective maintenance is really redesign.
Mean time to repair has certain significance.	Mean time to repair has no significance.
Preventive maintenance is carried out to inhibit failures.	Preventive maintenance has no meaning whatsoever in software.
Hardware can be repaired by using spare modules.	Software failures cannot be repaired by using spare modules.

7.4 Fault Masking

This is the term used in fault-tolerant computing to state that a system having redundancy can tolerate a number of failures prior to its failure. More specifically, the implication of the term simply is that a problem has appeared somewhere within the digital system framework, but because of the nature of the design, the problem does not effect the overall operation of the system. The best known fault masking method is probably modular redundancy presented below [23].

7.4.1 Triple Modular Redundancy (TMR)

The triple modular redundancy (TMR) scheme was first proposed by Von Neumann [3] in 1956, in which three identical modules or units perform the same task simultaneously and the voter compares their outputs (*i. e.*, the modules') and sides with the majority. More specifically, the TMR system fails only when more than one module/unit fails or the voter fails. In other words, the TMR system can tolerate failure of a single unit or module. An important example of the TMR scheme application was the SATURN V launch vehicle computer, which used TMR with voters in the central processor and duplication in the main memory [24]. The block diagram of the TMR scheme is shown in Figure 7.1. Blocks in the diagram represent modules/units and the voter. In addition, the TMR system without the voter is inside the dotted rectangle. This system is basically a 2-out-of-3 identical unit system.

For independently failing units and the voter, the reliability of the system in Figure 7.1 is [23]

$$R_{tv} = \left(3R^2 - 2R^3\right) R_V \tag{7.1}$$

where

R_{tv} is the reliability of the TMR system with voter.
R is the unit or module reliability.
R_V is the voter reliability.

For the 100% reliable voter (*i. e.*, $R_V = 1$), Equation (7.1) becomes

$$R_{tv} = 3R^2 - 2R^3 \tag{7.2}$$

where

R_{tv} is the reliability of the TMR system with the perfect voter.

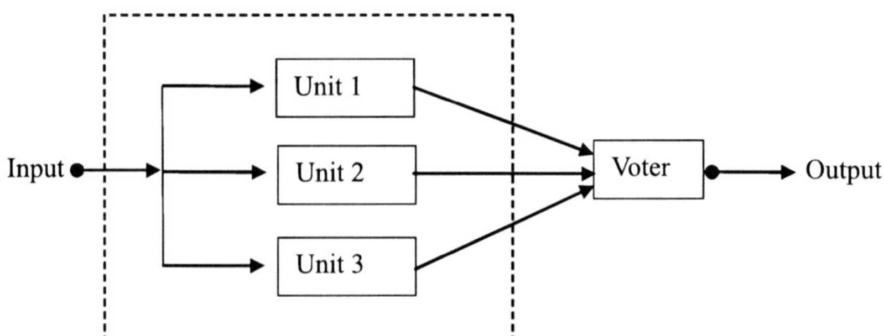

Figure 7.1. Block diagram representing the TMR scheme with voter

The voter reliability and the single unit's reliability determine the improvement in reliability of the TMR system over a single unit system. For the perfect voter (*i. e.*, $R_V = 1$), the TMR system reliability given by Equation (7.2) is only better than the single unit system when the reliability of the single unit is higher than 0.5.

At $R_V = 0.8$, the reliability of the TMR system is always less than the reliability of the single unit. Furthermore, when $R_V = 0.9$ the TMR system reliability is only marginally better than the single unit or module reliability when the single unit reliability is approximately between 0.667 and 0.833 [25].

- TMR System Maximum Reliability with Perfect Voter

For the perfect voter, the TMR system reliability is given by Equation (7.2). For this case, the ratio of R_{tv} to a single unit reliability, R, is given by [26]

$$\theta = \frac{R_{tv}}{R} = \frac{3R^2 - 2R^3}{R} = 3R - 2R^2 \tag{7.3}$$

Differetiating Equation (7.3) with respect to R and equating it to zero yields

$$\frac{d\theta}{dR} = 3 - 4R = 0 \tag{7.4}$$

By solving Equation (7.4), we get

$$R = 0.75$$

This simply means that the maximum reliability of the TMR system will occur at $R = 0.75$. Thus, inserting this value for R into Equation (7.2) yields

$$\begin{aligned} R_{tv} &= 3(0.75)^2 - 2(0.75)^3 \\ &= 0.8438 \end{aligned}$$

Thus, the maximum value of the TMR system reliability with the perfect voter is 0.8438.

Example 7.1

For a TMR system with a perfect voter, determine the points where the single unit and the TMR system reliabilities are equal.

In order to determine the points, we equate the single unit's reliability R with Equation (7.2) to get

$$R = R_{tv} = 3R^2 - 2R^3 \tag{7.5}$$

By rearranging Equation (7.5) we get

$$2R^2 - 3R + 1 = 0 \tag{7.6}$$

Obviously, Equation (7.6) is a quadratic equation and its roots are

$$R = \frac{3+[9-(4)(2)(1)]^{1/2}}{(2)(2)} = 1$$

and

$$R = \frac{3-[9-(4)(2)(1)]^{1/2}}{(2)(2)} = \frac{1}{2}$$

It means that the reliabilities of the TMR system and single unit are equal at $R = 1$, ½. The reliability of the TMR system will only be higher than the single unit reliability when the value of R is greater that 0.5.

- Mean Time to Failure of the TMR System

For constant failure rates of the TMR units and the voter, the TMR system with voter reliability using Equation (7.1) is given by [33]

$$\begin{aligned} R_{\text{tv}}(t) &= \left[3e^{-2\lambda t} - 2e^{-3\lambda t}\right]e^{-\lambda_v t} \\ &= 3e^{-(2\lambda+\lambda_v)t} - 2e^{-(3\lambda+\lambda_v)t} \end{aligned} \tag{7.7}$$

where

$R_{\text{tv}}(t)$ is the TMR system with voter reliability.
λ is the unit constant failure rate.
λ_{v} is the voter constant failure rate.

By integrating Equation (7.7) over the time interval from 0 to ∞, we obtain the following expression for the TMR system with voter mean time to failure [12]:

$$\begin{aligned} MTTF_{\text{tv}} &= \int_0^{\infty}\left[3e^{-(2\lambda+\lambda_v)t} - 2e^{-(3\lambda+\lambda_v)t}\right]dt \\ &= \frac{3}{(2\lambda+\lambda_v)} - \frac{2}{(3\lambda+\lambda_v)} \end{aligned} \tag{7.8}$$

where

$MTTF_{\text{tv}}$ is the mean time to failure of the TMR system with voter.

For $\lambda_{\text{v}} = 0$ or perfect voter, Equation (7.8) reduces to

$$\begin{aligned} MTTF_{\text{t}} &= \frac{3}{2\lambda} - \frac{2}{3\lambda} \\ &= \frac{5}{6\lambda} \end{aligned} \tag{7.9}$$

where

$MTTF_{\text{t}}$ is the TMR system with perfect voter mean time to failure.

Example 7.2
The constant failure rate of a unit belonging to a TMR system with voter is $\lambda = 0.0004$ failures per hour. Calculate the system reliability for a 500 hour mission if the voter constant failure rate is $\lambda_v = 0.00005$ failures per hour. In addition, calculate the system mean time to failure.

By using the specified data values in Equation (7.7), we get

$$R_{tv}(500) = 3e^{-[(2)(0.0004)+0.00005](500)} - 2e^{-[(3)(0.0004)+0.00005](500)}$$
$$= 0.8908$$

Similarly, inserting the given data values into Equation (7.8) yields

$$= \frac{3}{[(2)(0.0004)+0.00005]} - \frac{2}{[(3)(0.0004)+0.00005]}$$
$$= 1929.4 \text{ hours}$$

Thus, the TMR system reliability and mean time to failure are 0.8908 and 1929.4 hours, respectively.

7.4.2 *N*-Modular Redundancy (NMR)

This is the general form of the TMR (*i. e.*, it contains N identical units instead of only three). The number N is any odd number, and the NMR system can tolerate a maximum of m unit/modular failures if the value of N is equal to $(2m+1)$.

For independently failing units and the voter, the reliability of the NMR system with voter is given by [12]:

$$R_{NV} = R_V\left[\sum_{j=0}^{m}\binom{N}{j}R^{N-j}(1-R)^j\right] \tag{7.10}$$

where

$$\binom{N}{j} \equiv \frac{N!}{(N-j)!\,j!}$$

R_{NV} is the NMR system with voter reliability.
R is the unit/module reliability.
R_V is the voter reliability.

There are many other redundancy schemes used in computers. Some of these are described in Ref. [12].

7.5 Computer System Life Cycle Costing

The life cycle costing concept is often used in the industrial sector, especially in the procurement of expensive items [27]. In regard to computers, the life cycle cost of a computer may simply be defined as the total of all costs to buyers (*i. e.*, the costs associated with procurement and ownership of the computer) over its entire life span. Some of the uses of the life cycle costing concept with regard to computer systems can be as follows [12]:

- To choose a manufacturer of a computer system out of many competing manufacturers
- To make effective decisions for computer system replacement
- To compare the costs of alternative approaches to meet a requirement

This section presents three mathematical models: one to estimate computer system life cycle cost, and the other two to estimate computer system ownership-related costs only.

Model I

Model I is concerned with estimating the computer system life cycle cost. The life cycle cost of a computer system is expressed by

$$LCC_{\mathrm{CS}} = CSPC + CSOC \tag{7.11}$$

where

LCC_{CS} is the life cycle cost of a computer system.
$CSPC$ is the computer system procurement cost.
$CSOC$ is the computer system ownership cost.

Model II

Model II is concerned with estimating annual labour cost associated with servicing a computer system. Thus, the annual labour cost is expressed by

$$ASCC = (AH)(HLC)(\theta_1 + \theta_2) \tag{7.12}$$

where

$$\theta_1 = \frac{(MTTR + TT)}{MTBF} \tag{7.13}$$

$$\theta_2 = \frac{(ATPPM + TT_{\mathrm{pm}})}{ATBPM} \tag{7.14}$$

The symbols used in Equations (7.12)–(7.14) are defined below.

$ASCC$ is the annual labour cost associated with servicing a computer system.
AH is the number of hours in one year (*i. e.*, 8,760 hours).
HLC is the hourly labour cost.
$MTTR$ is the mean time to repair of the computer system.
TT is the travel time associated with a repair call.
$MTBF$ is the mean time between failures of the computer system.
$ATPPM$ is the average time to perform preventive maintenance.
TT_{pm} is the travel time associated with a preventive maintenance call.
$ATBPM$ is the average time between preventive maintenance.

Model III

Model III is concerned with estimating the monthly maintenance cost of computer system hardware. Thus, the computer system hardware monthly maintenance cost is expressed by [28]:

$$CSHMC = PMC + CMC + IC \tag{7.15}$$

where

CSHMC is the computer system hardware maintenance cost per month.
PMC is the preventive maintenance cost per month.
IC is the cost of inventory per month.
CMC is the corrective maintenance cost per month.

The costs *PMC*, *CMC*, and *IC* are defined as follows:

$$PMC = \frac{(OH)(HR)[CETPM + TTCEPM]}{SPMI} \tag{7.16}$$

$$CMC = \frac{(OH)(HR)[MTTR + TTCECM]}{MTBF} \tag{7.17}$$

$$IC = (ICR)(MSPOMC) \tag{7.18}$$

where

OH is the equipment operating hours per month.
HR is the hourly rate of the customer engineer.
CETPM is the customer engineer's scheduled time for performing preventive maintenance.
TTCEPM is the travel time of the customer engineer for performing preventive maintenance.
SPMI is the scheduled preventive maintenance interval.
MTTR is the mean time to repair.
TTCECM is the customer engineer's travel time for corrective maintenance.
MTBF is the mean time between failures.
ICR is the monthly inventory cost rate (This includes monthly handling costs and spares' depreciative charges).
MSPOMC is the maintenance spare parts original manufacturing cost (*i. e.*, inventory value).

The customer engineer's hourly rate is given by

$$HR = PHC + \left[CEHP\left(1+i\right)\right]/\alpha \tag{7.19}$$

where

PHC is the parts' hourly cost.
i is the overhead rate.
CEHP is the customer engineer's hourly pay.
α is the fraction of the total time customer engineer spends for the maintenance purpose.

7.6 Software Reliability Evaluation Models

Over the years many mathematical models to evaluate software reliability have been developed [5, 29, 30]. This section presents two such models.

7.6.1 Mills Model

This model was developed by H.D. Mills in 1972 by arguing that the faults remaining in a given software program can be estimated through a seeding process that makes an assumption of a homogeneous distribution of a representative class of faults [31]. Thus, both seeded and unseeded faults are identified during reviews or testing and the discovery of seeded and unseeded faults permits an assessment of remaining faults for the fault type in question.

The maximum likelihood of the unseeded faults is defined by [25]

$$M_u = [M_s n_u] / n_s \tag{7.20}$$

where

M_u is the maximum likelihood of the unseeded faults.
M_s is the total number of seeded faults.
n_u is the total number of unseeded faults uncovered.
n_s is the total number of seeded faults found.

Thus, the number of unseeded faults still remaining in a program under consideration is

$$M = M_u - n_u \tag{7.21}$$

Example 7.3
A software program was seeded with a total of 20 faults and, during testing, 45 faults of the same kind were discovered. Fifteen of these faults were the seeded faults and the remaining thirty unseeded faults.

Estimate the number of unseeded faults still remaining in the program.

By substituting the specified data values into Equation (7.20), we get

$$M_u = \frac{(20)(30)}{15} = 40 \text{ faults}$$

Using the above calculated value and the other given data value in Equation (7.21) yields

$$M = 40 - 30 = 10 \text{ faults}$$

It means ten unseeded faults still remain in the program.

7.6.2 Musa Model

This model is based on the premise that reliability assessments in the time domain can only be based upon actual or real execution time, as opposed to elapsed or calendar time, because only during execution a software program really becomes exposed to failure-provoking stress. Some of the important assumptions pertaining to this model are as follows [12, 23, 32]:

- Failure intervals follow a Poisson distribution and are statistically independent.
- Failure rate is proportional to the remaining defects.
- Execution times between failures are piecewise exponentially distributed.

A comprehensive list of assumptions is available in Ref. [12]. The net number of corrected faults is expressed by [23, 32]:

$$n = N\left[1 - \exp\left(-\alpha t / N T_m\right)\right] \tag{7.22}$$

where

n is the net number of corrected software faults.

t is time.

α is the testing compression factor and is defined as the average ratio of detection rate of failures during test to the rate during normal application of the software program.

T_m is the mean time to failure at the start of the test.

N is the number of initial faults.

Mean time to failure, T, increases exponentially with execution time and is defined by

$$T = T_m \exp\left(\alpha t / N T_m\right) \tag{7.23}$$

Thus, the reliability at operational time t is expressed by

$$R(t) = \exp\left(-t / T\right) \tag{7.24}$$

From the above relationships, the number of failures that must occur to increase mean time to failure from, say, T_a to T_b [33]:

$$\Delta n = N T_m \left[\frac{1}{T_a} - \frac{1}{T_b}\right] \tag{7.25}$$

The additional execution time needed to experience Δn is expressed by

$$\Delta n = \left[\frac{N T_m}{\alpha}\right] \ln\left(\frac{T_b}{T_a}\right) \tag{7.26}$$

Example 7.4
Assume that a software program is estimated to have around 500 errors, and at the start of the testing process the recorded mean time to failure is 5 hours.

Estimate the test time required to reduce the remaining errors to 20, if the value of the testing compression factor is 6. Calculate reliability over a 50-hour operational period.

Using the given data values in Equation (7.25) yields

$$(500-20)=(500)(5)\left[\frac{1}{5}-\frac{1}{T_b}\right] \tag{7.27}$$

By rearranging Equation (7.27) we get

$$T_b = 125 \text{ hours}$$

By substituting the above calculated value and the other given data values into Equation (7.26), we get

$$\Delta t = \left[\frac{(500)(5)}{6}\right]\ln\left(\frac{125}{5}\right)$$
$$= 1{,}341.2 \text{ hours}$$

Similarly, using the calculated and given data values in Equation (7.24) yields

$$R(50) = \exp\left(-\frac{50}{125}\right)$$
$$= 0.6703$$

Thus, the test time required to reduce errors to 20 is 1,341.2 hours and the software reliability for the given operational period is 0.6703.

7.7 Internet Reliability, Failure Examples, Outage Categories, and Related Observations

The demand for Internet reliability continues to escalate as the Internet evolves to support various applications including telephony and banking [34]. However, various studies conducted over the past decade indicate that the reliability of Internet paths falls far short of the 99.999% availability expected in the public-switched telephone network (PSTN) [35]. Furthermore, small-scale studies conducted in 1994 and 2000 revealed that the probability of encountering a major routing pathology along a path is approximately 1.5% to 3.3% [34, 36]. Over the years various means have been used to improve Internet reliability, including server replication, multi-homing, and overlay networks [34].

Some examples of the Internet failures are as follows [37]:

- On August 14, 1998, a misconfigured important Internet database server mistakenly referred all queries for Internet machines with names ending in ".net" to the incorrect secondary database server. As the result of this problem, most of connections to ".net" Internet Web servers and other end stations failed for many hours [38].
- On April 25, 1997, a misconfigured router of a Virginia service provider injected a wrong map into the global Internet. In turn, the Internet providers that accepted this incorrect map automatically diverted their concerned traffic to the Virginia provider. This resulted in network congestion, instability, and overload of Internet router table memory that ultimately shut down most of the major Internet backbones for up to two hours [39].
- On November 8, 1998, a malformed routing control message caused by a software fault triggered an interoperability problem between various core Internet backbone routers produced by different vendors. This resulted in a wide-spread loss of network connectivity (*i. e.*, experienced by the Internet end-users) as well as increment in packet loss and latency. All in all, it took several hours for majority of backbone providers to resolve this outage effectively [40].

A case study conducted over a period of one year (*i. e.*, November 1997 to November 1998) concerning Internet outages classified the outages into many categories (along with their occurrence percentages in parentheses) as follows [37]:

- Maintenance (16.2%)
- Power outage (16%)
- Fiber cut/circuit/carrier problem (15.3%)
- Unreachable (12.6%)
- Hardware problem (9%)
- Interface down (6.2%)
- Routing problem (6.1%)
- Miscellaneous (5.9%)
- Unknown/undetermined/no problem (5.6%)
- Congestion/sluggish (4.6%)
- Malicious attacks (1.5%)
- Software problem (1.3%)

As per the findings of one study, some of the Internet reliability-related observations are as follows [37]:

- Availability and mean time to failure of the Internet backbone infrastructure are significantly less than the Public Switched Telephone Network (PSTN).
- There is only a small fraction of network paths in the Internet infrastructure that contribute disproportionately to the number of long-term outages and backbone unavailability.

- The most of Internet backbone paths exhibit a mean time to failure of about 25 days or les and a mean time to repair of around twenty minutes or less.
- It appears that most inter-provider path failures result from congestion collapse.

7.8 An Approach for Automating Fault Detection in Internet Services

As most Internet services (*e. g.*, search engines and e-commerce) suffer faults, a quick detection of these faults could be an important factor in improving the availability of the system. This approach known as the pinpoint method combines the easy deploy-ability of low-level monitors with the higher-level monitors' ability for detecting application-level faults [41]. This method is based upon the following assumptions with respect to the system under observation and its workload [41]:

- The software is made up of various interconnected components (modules) with well-defined narrow interfaces. These could be software subsystems, objects, or simply physical mode boundaries.
- There is a high volume of basically independent requests (*i. e.*, from different users).
- An interaction with the system is relatively short-lived, whose processing can be broken down as a path. More specifically, a tree of the names of elements or components that take part in the servicing of that request.

The pinpoint approach to detecting and localizing anomalies is basically a three stage process [41]:

- **Observing the system.** This is concerned with capturing the runtime path of each request served by the system and then from these paths extracting two specific low-level behaviours likely to reflect high-level functionality: path shapes and interactions of components.
- **Learning the patterns in system behaviour.** This is concerned with constructing a reference model representing the normal behaviour of an application in regard to path shapes and component interactions, by assuming that most of the system functions correctly most of the time.
- **Detecting anomalies in system behaviours.** This is concerned with analyzing the system's current behaviour and detecting anomalies with respect to the reference model.

The pinpoint approach is described in detail in Ref. [41].

7.9 Internet Reliability Models

There are many mathematical models that can be used to perform reliability-related analysis in various areas of Internet [42–45]. This section presents two of these models.

7.9.1 Model I

Model I is concerned with evaluating the reliability and availability of a server system. The model assumes that the Internet server system can either be in an operating or a failed state and its failure/outage and restoration/repair rates are constant. The server system state space diagram is shown in Figure 7.2. The numerals in boxes denote the system state.

Using the Markov method, we write down the following two differential equations for Figure 7.2 state space diagram [23]:

$$\frac{dP_0(t)}{dt} + \lambda_s P_0(t) = \mu_s P_1(t) \tag{7.28}$$

$$\frac{dP_1(t)}{dt} + \mu_s P_1(t) = \lambda_s P_0(t) \tag{7.29}$$

At time $t=0$, $P_0(0)=1$ and $P_1(0)=0$.

The symbols used in Equations (7.28) and (7.29) are defined below.

$P_i(t)$ is the probability that the server system is in state i at time t, for $i=0, 1$.
λ_s is the server system constant failure/outage rate.
μ_s is the server system constant repair/restoration rate.

Solving Equations (7.28) and (7.29), we get [23]

$$P_0(t) = A_s(t) = \frac{\mu_s}{(\lambda_s + \mu_s)} + \frac{\lambda_s}{(\lambda_s + \mu_s)} e^{-(\lambda_s + \mu_s)t} \tag{7.30}$$

$$P_1(t) = UA_s(t) = \frac{\lambda_s}{(\lambda_s + \mu_s)} - \frac{\lambda_s}{(\lambda_s + \mu_s)} e^{-(\lambda_s + \mu_s)t} \tag{7.31}$$

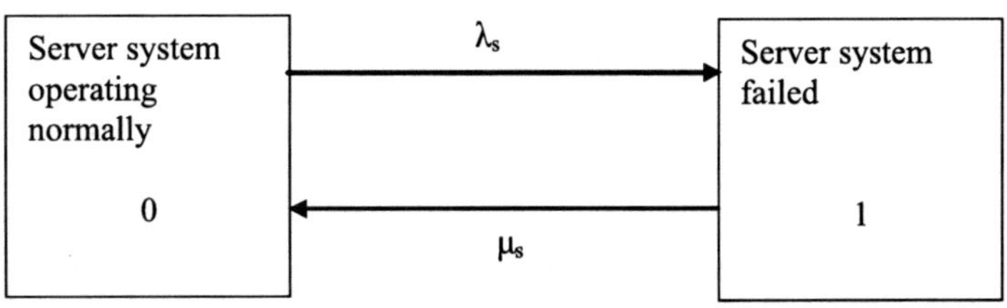

Figure 7.2. Server system transition diagram

where

$A_s(t)$ is the server system availability at time t.

$UA_s(t)$ is the server system unavailability at time t.

As time t becomes very large, Equations (7.30) and (7.31) reduce to

$$A_s = \frac{\mu_s}{\lambda_s + \mu_s} \tag{7.32}$$

and

$$UA_s = \frac{\lambda_s}{\lambda_s + \mu_s} \tag{7.33}$$

where

A_s is the server system steady state availability.

UA_s is the server system steady state unavailability.

For $\mu_s = 0$, Equation (7.30) reduces to

$$R_s(t) = e^{-\lambda_s t} \tag{7.34}$$

where

$R_s(t)$ is the server system reliability at time t.

Thus, the server system mean time to failure ($MTTF_s$) is given by [23]

$$\begin{aligned} MTTF_s &= \int_0^\infty R_s(t)\,dt \\ &= \int_0^\infty e^{-\lambda_s t}\,dt \\ &= \frac{1}{\lambda_s} \end{aligned} \tag{7.35}$$

Example 7.5

Assume that the constant outage and restoration rates of an Internet server system are 0.0045 outages/hour and 0.05 restorations/hour, respectively. Calculate the server system steady state availability.

By substituting the given data values into Equation (7.32) we get

$$\begin{aligned} A_s &= \frac{0.05}{0.0045 + 0.05} \\ &= 0.9174 \end{aligned}$$

Thus, the steady state availability of the server system is 0.9174.

7.9.2 Model II

Model II is concerned with evaluating the availability of an Internetworking (router) system composed of two independent and identical switches. The model assumes that the system fails when both the switches fails and the switches form a standby-type configuration. In addition, the switch failure and restoration rates are constant. The system state space diagram is shown in Figure 7.3. The numerals in circles denote the system state.

Using the Markov method, we write down the following differential equations for Figure 7.3 diagram [23,46]:

$$\frac{dP_0(t)}{dt}+\left[p\lambda_{sw}+(1-p)\lambda_{sw}\right]P_0(t)=\mu_{sw}P_1(t)+\mu_{sw1}P_2(t) \quad (7.36)$$

$$\frac{dP_1(t)}{dt}+(\lambda_{sw}+\mu_{sw})P_1(t)=p\lambda_{sw}P_0(t) \quad (7.37)$$

$$\frac{dP_2(t)}{dt}+\mu_{sw1}P_2(t)=\lambda_{sw}P_1(t)+(1-p)\lambda_{sw}P_0(t) \quad (7.38)$$

At time $t=0$, $P_0(0)=1$ and $P_1(0)=P_2(0)=0$.

The symbols used in Equations (7.36)–(7.38) are defined below.

$P_i(t)$ is the probability that the Internetworking (router) system is in state i at time t, for $i=0, 1, 2$.

λ_{sw} is the switch constant failure rate.

p is the failure detection and successful switchover probability from switch failure.

μ_{sw} is the switch constant repair/restoration rate.

μ_{sw1} is the constant restoration/repair rate from system state 2 to state 0.

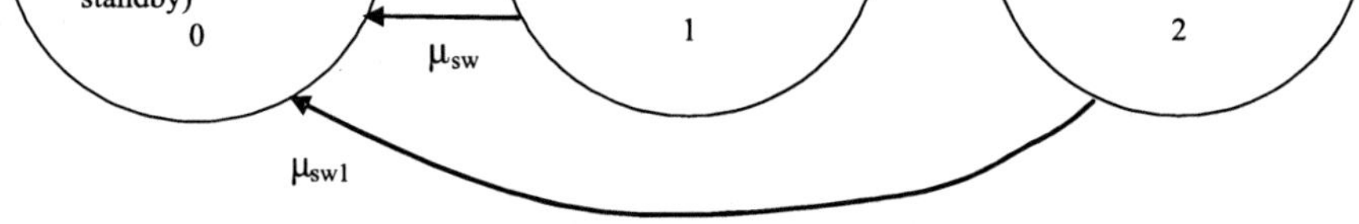

Figure 7.3. System transition diagram

The following steady-state probability solutions are obtained by setting derivatives equal to zero in Equations (7.36)–(7.38) and using the relationship $\sum_{i=0}^{2} P_i = 1$.

$$P_0 = \frac{\mu_{\text{sw}1}(\mu_{\text{sw}} + \lambda_{\text{sw}})}{A} \tag{7.39}$$

$$P_1 = \frac{p\lambda_{\text{sw}}\,\mu_{\text{sw}1}}{A} \tag{7.40}$$

$$P_2 = \frac{p\lambda_{\text{sw}}^2 + (1-p)\lambda_{\text{sw}}(\mu_{\text{sw}} + \lambda_{\text{sw}})}{A} \tag{7.41}$$

where

$$A \equiv \mu_{\text{sw}1}\left(\mu_{\text{sw}} + p\lambda_{\text{sw}} + \lambda_{\text{sw}}\right) + (1-p)\lambda_{\text{sw}}\left(\mu_{\text{sw}} + \lambda_{\text{sw}}\right) + p\lambda_{\text{sw}}^2 \tag{7.42}$$

P_i is the steady state probability that the Internetworking (Router) system is in state i, for $i = 0, 1, 2$.

The system steady state availability is given by

$$\begin{aligned} A_s &= P_0 + P_1 \\ &= \frac{\mu_{\text{sw}1}\left(\mu_{\text{sw}} + \lambda_{\text{sw}} + p\lambda_{\text{sw}}\right)}{A} \end{aligned} \tag{7.43}$$

7.10 Problems

1. Write an essay on developments in computer hardware and software reliability.
2. What are the main causes for computer system failures?
3. Make a comparison between hardware and software reliability.
4. What is fault masking?
5. Assume that the constant failure rate of a unit belonging to a TMR system with voter is $\lambda = 0.0005$ failures per hour. Calculate the system reliability for a 400 hour mission if the voter constant failure rate is $\lambda_V = 0.00001$ failures per hour. In addition, calculate the system mean time to failure.
6. A software program was seeded with 25 faults and, during testing, 50 faults of the same type were found. Twenty of these faults were the seeded faults and the remaining thirty unseeded faults. Calculate the number of unseeded faults still remaining in the program.
7. Compare the Musa model with the Mills model.
8. Discuss Internet failures and their consequences.
9. Describe a method for automating fault detection in Internet services.
10. Prove Equations (7.39)–(7.41).

References

1 Shannon, C.E., A Mathematical Theory of Communications, Bell System Tech. J., Vol. 27, 1948, pp. 379–423 and 623–656.
2 Hamming, W.R., Error Detecting and Error Correcting Codes, Bell System Tech. J., Vol. 29, 1950, pp. 147–160.
3 Von Neumann, J., Probabilistic Logics and the Synthesis of Reliable Organisms from Reliable Components, in Automata Studies, edited by C.E., Shannon and J. McCarthy, Princton University Press, Princeton, New Jersey, 1956, pp. 43–48.
4 Moore, E.F., Shannon, C.E., Reliable Circuits Using Less Reliable Relays, J. Franklin Inst., Vol. 262, 1956, pp. 191–208.
5 Schick, G.J., Wolverton, R.W., A Analysis of Competing Software Reliability Models, IEEE Trans. on Software Engineering, Vol. 4, 1978, pp. 140–145.
6 Haugk, G., Tsiang, S.H., Zimmermann, L., System Testing of the No. 1 Electronic Switching System, Bell System Tech. J., Vol. 43, 1964, pp. 2575–2592.
7 Floyd, R.W., Assigning Meanings to Program, Math. Aspects Comp. Sci., Vol. XIX, 1967, pp. 19–32.
8 Hudson, G.R., Programming Errors as a Birth-and-Death Process, Report No. SP-3011, System Development Corporation, 1967.
9 Barlow, R., Scheuer, E.M., Reliability Growth During a Development Testing Program, Technometrics, Vol. 8, 1966, pp. 53–60.
10 London, R.L., Proving Program Correct: Some Techniques and Examples, BIT, Vol. 10, 1969, pp. 168–182.
11 Sauter, J.L., Reliability in Computer Programs, Mechanical Engineering, Vol. 91, 1969, pp. 24–27.
12 Dhillon, B.S., Reliability in Computer System Design, Ablex Publishing, Norwood, New Jersey, 1987.
13 Goseva-Popstojanova, K., Mazimdar, S., Singh, A.D., Empirical Study of Session-Based Workload and Reliability for Web Servers, Proceedings of the 15th Int. Symposium on Software Reliability Engineering, 2004, pp. 403–414.
14 Yourdon, E., The Causes of System Failures – Part 2, Modern Data, Vol. 5, Feb. 1972, pp. 50–56.
15 Yourdon, E., The Causes of System Failures – Part 3, Modern Data, Vol. 5, March 1972, pp. 36–40.
16 Bouricious, W.G., Carter, W.C., Schneider, P.R., Reliability Modelling Techniques for Self-Repairing Computer Systems, Proceedings of the 12th Association of Computing Machinery National Conference, 1969, pp. 295–305.
17 Mathur, F.P., Avizienis, A., Reliability Analysis and Architecture of a Highly Redundant Digital System: Generalized Triple Modular Redundancy with Self-Repair, Proceedings of the American Federation of Information Processing Societies (AFIPS), 1970, pp. 375–383.
18 Losq, J., A Highly Efficient Redundancy Scheme: Self-Purging Redundancy, IEEE Transactions on Computers, Vol. 25, June 1976, pp. 569–578.
19 Borgerson, B.R., Freitas, R.F., A Reliability Model for Gracefully Degrading Standby-Sparing Systems, IEEE Transactions on Computers, Vol. 24, 1975, pp. 517–525.

20 Kline, M.B., Software and Hardware Reliability and Maintainability: What are the Differences? Proceedings of the Annual Reliability and Maintainability Symposium, 1980, pp. 179–185.
21 Grant Ireson, W., Coombs, C.F., Moss, R.Y., Handbook of Reliability Engineering and Management, McGraw Hill Book Company, New York, 1996.
22 Dhillon, B.S., Reliability Engineering in Systems Design and Operation, Van Nostrand Reinhold Company, New York, 1983.
23 Dhillon, B.S., Design Reliability: Fundamentals and Applications, CRC Press, Boca Raton, Florida, 1999.
24 Mathur, F.P., Avizienis, A., Reliability Analysis and Architecture of a Hybrid Redundant Digital System: Generalized Triple Modular Redundancy with Self-Repair, Proceedings of the AFIPS Spring Joint Computer Conference, 1970, pp. 375–387.
25 Pecht, M., Editor, Product Reliability, Maintainability, and Supportability Handbook, CRC Press, Boca Raton, Florida, 1995.
26 Shooman, M.L., Fault-Tolerant Computing, Annual Reliability and Maintainability Symposium Tutorial Notes, 1994, pp. 1–25.
27 Dhillon, B.S., Life Cycle Costing: Techniques, Models, and Applications, Gordon and Breach Science Publishers, New York, 1989.
28 Phister, M., Data Processing Technology and Economics, Santa Monica Publishing Company, Santa Monica, California, 1979.
29 Musa, J.D., Iannino, A., Okumoto, K., Software Reliability, McGraw-Hill Book Company, New York, 1987.
30 Sukert, A.N., An Investigation of Software Reliability Models, Proceedings of the Annual Reliability and Maintainability Symposium, 1977, pp. 478–484.
31 Mills, H.D., On the Statistical Validation of Computer Programs, Report No. 72-6015, 1972. IBM Federal Systems Division, Gaithersburg, Maryland, U.S.A.
32 Musa, J.D., A Theory of Software Reliability and Its Applications, IEEE Transactions on Software Engineering, Vol. 1, 1975, pp. 312–327.
33 Dunn, R., Ullman, R., Quality Assurance for Computer Software, McGraw-Hill Book Company, New York, 1982.
34 Gummadi, K.P., et al., Improving the Reliability of Internet Paths with One-Hop Source Routing, Proceedings of the 6th Usenix/ACM Symposium on Operating Systems and Design Implementation (OSDI), 2004, pp. 183–198.
35 Kuhn, D.R., Sources of Failure in the Public Switched Telephone Networks, IEEE Transactions on Computers, Vol. 30, No. 4, 1997, pp. 31–36.
36 Paxson, V., End-to-End Routing Behaviour in the Internet, IEEE/ACM Transactions on Networking, Vol. 5, No. 5, 1997, pp. 601–615.
37 Lapovitz, C., Ahuja, A., Jahamian, F., Experimental Study of Internet Stability and Wide-Area Backbone Failures, Proceedings of the 29th Annual International Symposium on Fault-Tolerant Computing, 1999, pp. 278–285.
38 North American Network Operators Group (NANOG) mailing list, http://www.merit.edu/mail.archives/html/nanog/msg00569.html.
39 Barrett, R., Haar, S., Whitestone, R., Routing Snafu Causes Internet Outage, Interactive Week, April 25, 1997.
40 North American Network Operators Group (NANOG) mailing list, http://www.merit.edu/mail.archives/html/nanog/msg03039.html.

41 Kiciman, E., Fox, A., Detecting Application-Level Failures in Component-Based Internet Services, IEEE Transactions on Neural Networks, Vol. 16, No. 5, 2005, pp. 1027–1041.
42 Hecht, M., Reliability/Availability Modeling and Prediction for E-Commerce and Other Internet Information Systems, Proceedings of the Annual Reliability and Maintainability Symposium, 2001, pp. 176–182.
43 Aida, M., Abe, T., Stochastic Model of Internet Access Patterns, IEICE Transactions on Communications, Vol. E 84 – B, No. 8, 2001, pp. 2142–2150.
44 Chan, C.K., Tortorella, M., Spares-Inventory Sizing for End-to-End Service Availability, Proceedings of the Annual Reliability and Maintainability Symposium, 2001, pp. 98–102.
45 Imaizumi, M., Kimura, M., Yasui, K., Optimal Monitoring Policy for Server System With Illegal Access, Proceedings of the 11th ISSAT International Conference on Reliability and Quality in Design, 2005, pp. 155–159.
46 Dhillon, B.S., Kirmizi, F., Probabilistic Safety Analysis of Maintainable Systems, Journal of Quality in Maintenance Engineering, Vol. 9, No. 3, 2003, pp. 303–320.

8

Quality in Health Care

8.1 Introduction

Each year billions of dollars are being spent on health care worldwide. For example, in 1992 the United States spent $840 billion on health care, or 14% of its gross domestic product (GDP) [1]. Furthermore, since 1960 the health care spending in the United States has increased from 5.3% of the gross national product (GNP) to 13% in 1991 [2].

The history of quality in health care may be traced back to the 1860s, when Florence Nightingale (1820–1910), a British nurse, helped to lay the foundation for the health care quality assurance programs, by advocating the need for a uniform system for the collection and evaluation of hospital-related statistics [1]. Her analysis of the data collected showed that mortality rates varied quite significantly from one hospital to another.

In 1914, in the Untied States E.A. Codman (1869–1940) studied the results of health care with respect to quality, and emphasized the issues, when examining the quality of care, such as the accreditation of institutions, the importance of licensure or certification of providers, the need for taking into consideration the severity or stage of the disease, the economic barriers to receiving care, and the health and illness behaviours of the patients [1, 3].

Over the years, many other people have contributed to the field of quality in health care. An extensive list of publications on the topic is presented at the end of this book. This chapter presents various important aspects of quality in health care.

8.2 Health Care Quality Terms and Definitions and Reasons for the Rising Health Care Cost

Some of the commonly used terms and definitions in health care quality are as follows [4, 5]:

- **Health care.** This is services provided to individuals or communities for promoting, maintaining, monitoring, or restoring health.
- **Quality.** This is the extent to which the properties of a product or service generate/produce a desired outcome.
- **Quality assurance.** This is the measurement of the degree of care given (assessment) and, when appropriate, mechanisms for improving it.
- **Total quality management.** This is a philosophy of pursuing continuous improvement in each and every process through the integrated efforts of all concerned individuals associated with the organization.
- **Quality of care.** This is the level to which delivered health services satisfy established professional standards and judgements of value to consumers.
- **Quality improvement.** This is the total of all the appropriate activities that create a desired change in quality.
- **Clinical audit.** This is the process of reviewing the delivery of care against established standards to identify and remedy all deficiencies through a process of continuous quality improvement.
- **Cost of quality.** This is the expense of not doing effectively all the right things right the first time.
- **Quality assessment.** This is the measurement of the degree of quality at some point in time, without any effort for improving or changing the degree of care.
- **Dimensions of quality.** These are the measures of health system performance, including measures of effectiveness, appropriateness, efficiency, safety, continuity, accessibility, capability, sustainability, and responsiveness.
- **Adverse event.** This is an incident in which unintended harm resulted to an individual receiving health care.

There are many reasons for the rising health care cost. Some of these are shown in Figure 8.1 [6]. Each of these reasons is discussed in detail in Refs. [2, 6].

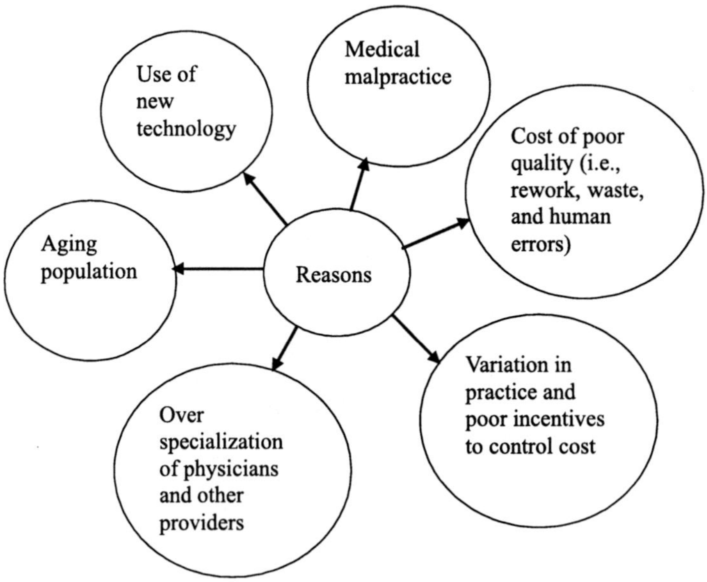

Figure 8.1. Some of the main reasons for the escalating health care cost

8.3 Comparisons of Traditional Quality Assurance and Total Quality Management with Respect to Health Care and Quality Assurance Versus Quality Improvement in Health Care Institutions

A comparison of traditional quality assurance and total quality management, directly or indirectly, with respect to many different areas of health care is presented in Table 8.1 [2].

Over years various authors have discussed the differences between quality assurance and quality improvement in health care institutions [7–11]. A clear understanding of these differences is important, as they contribute to differing information needs. Most of these differences are presented in Table 8.2 [7–11].

Table 8.1. Comprisons of traditional quality assurance and total quality management with respect to health care

No.	Area (characteristic)	Traditional quality assurance	Total quality management
1	Purpose	Enhance quality of patient care for patients	Enhance all products and services quality for patients and other customers
2	Aim	Problem solving	Continuous improvement, even when no deficiency/ problem is identified
3	Leadership	Physician and clinical leaders (*i. e.*, clinical staff chief and quality assurance committee)	All leaders (*i. e.*, clinical and non-clinical)
4	Customer	Customers are review organizations and professionals with focus on patients	Customers are review organizations, patients, professionals, and others
5	Scope	Clinical processes and outcomes	All processes and systems (*i. e.*, clinical and non-clinical)
6	Focus	Peer review vertically focused by clinical process or department (*i. e.*, each department looks after its own quality assurance)	Horizontally focused peer review for improving all processes and individuals that affect outcomes
7	People involved	Appointed committees and quality assurance program	Each and every person involved with process
8	Methods	Includes hypothesis testing, chart audits, indicator monitoring, and nominal group techniques	Includes checklist, force field analysis, Pareto chart, indicator monitoring and data use, Hoshin planning, brainstorming, flowcharts, nominal group techniques, quality function deployment, control chart, fishbone diagram, etc.
9	Outcomes	Includes measurement and monitoring	Includes also measurement and monitoring

Table 8.2. Comparisons of quality assurance and quality improvement in health care institutions

No.	Area (characteristic)	Quality improvement	Quality assurance
1	Goal	Satisfy customer requirements	Regulatory compliance
2	Participants	Every associated person	Peers
3	Viewpoint	Proactive	Reactive
4	Focus	All involved processes	Physician
5	Review technique	Analysis	Summary
6	Customers	Patients, caregivers, payers, technicians, enrollees, support staff, managers, etc.	Regulators
7	Performance measure	Need/capability	External standards
8	Direction	Decentralized through the management line of authority	Committee or central coordinator
9	Functions involved	Many (clinician and support system)	Few (mainly doctors)
10	Action taken	Implement appropriate improvements	Recommend appropriate improvements
11	Defects studied	Special and common causes	Outliers special causes

8.4 Assumptions Guiding the Development of Quality Strategies in Health Care, Health Care-related Quality Goals and Strategies, Steps for Quality Improvement, and Physician Reactions to Total Quality

A clear understanding of the assumptions guiding the development of quality strategies in health care is necessary for the ultimate success of these strategies. Some of these assumptions are [1]:

- Total quality management is an important unifying leadership philosophy that encompasses all functions of a health care organization, not just the quality assurance function and clinical care.
- The measurement of quality care must include items such as the determination of patient outcomes, patient feedback and involvement, cost effectiveness, assurance of appropriateness of care, review of key internal processes, and proper coordination of care across a continuum of services and providers.
- Total quality management (TQM) is a good means of furthering the organizational culture and mission. More specifically, this basically means that quality results from continuously improving care and work processes, patients and others served are the highest priority and should have a rather strong voice in the

design and delivery of care, quality must flow from leadership and permeate all levels of the organization, decisions should be based on facts, but reflect compassion and caring, and processes are improved by teamwork and involvement.

- The system will be increasingly responsible for delivering the quality of care to all enrolled people on a regional basis.
- Quality improvement definitely needs timely access to reliable clinical data and an effective capability for analyzing and interpreting clinical pathways.

Four important health care-related quality goals are shown in Figure 8.2 [1]. Three useful strategies associated with Goal I are as follows [1]:

- Aim to maximize patients' and families' involvement in the care experience by using shared decision making and improving patient involvement in care choices.
- Ensure, in an effective manner, the assessment of employee, patient, and medical staff satisfaction periodically by incorporating survey standards and benchmarking.
- Implement recommendations concerning compassionate care of dying and carefully address the spiritual needs of patients and families through pastoral care.

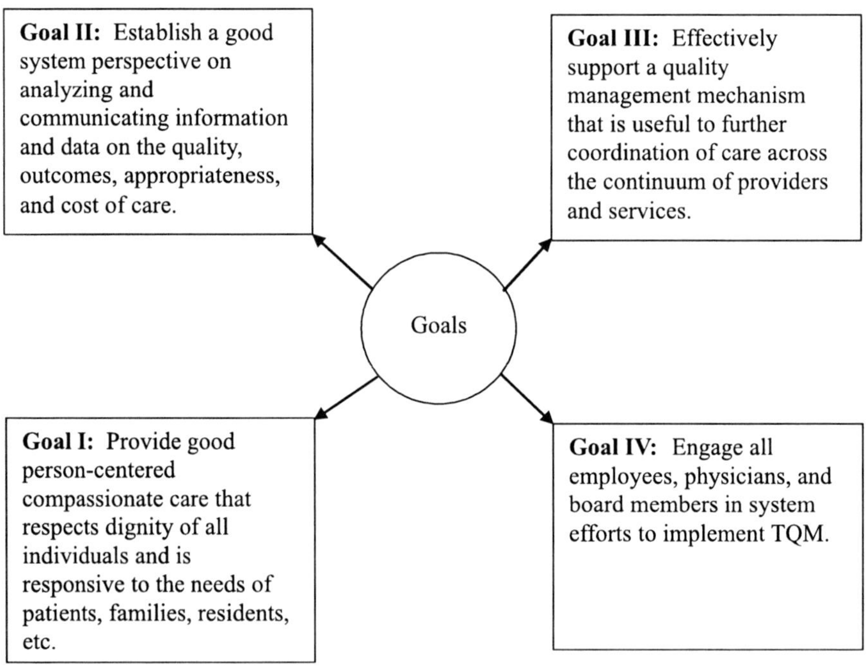

Figure 8.2. Health care-related quality goals

Three strategies pertaining to Goal II are as follows [1]:

- Establish a system plan for addressing information needs concerning quality management, including a pivotal clinical data set, common definitions, and enhanced analysis of available information.
- Document and share critical quality performance and outcome studies throughout the system and assess the implications of new developments in the evolution of electronic medical records.
- Further develop the competencies and skills of individuals associated with quality through user conferences and other appropriate means.

Two of the strategies concerning Goal III are as follows [1]:

- Develop further and apply case management models across the continuum of services.
- Determine the ways the development of integrated delivery systems can help to promote access and quality of care.

Three strategies associated with Goal IV are as follows [1]:

- Actively involve physicians when developing treatment protocols and improving care systems.
- Develop appropriate programs on TQM for people such as physicians, board members, and employees.
- Establish and apply appropriate management models that help to promote effective teamwork and participatory decision making.

Figure 8.3. presents ten steps that can be used in improving quality in the health care system [12].

There have been varying reactions to TQM by physicians over the years. Some of the typical ones are as follows [2]:

- TQM basically is quality assurance in different clothing.
- Physicians have always used the scientific method; thus the scientific method advocated by TQM is nothing new.
- The TQM concept is applicable to administrative systems and industrial processes, but not to the clinical care of patients.
- The application of the TQM concept will wrest control of the patient care process from physicians.
- The TQM concept is another cost-cutting mechanism by management that will limit access to resources physicians require for their patients.
- The application of the TQM concept is a further encroachment on the physician-patient relationships, as patient care cannot be standardized like industrial processes.
- The application of the TQM concept will lead to additional committee meetings for time-constrained physicians.

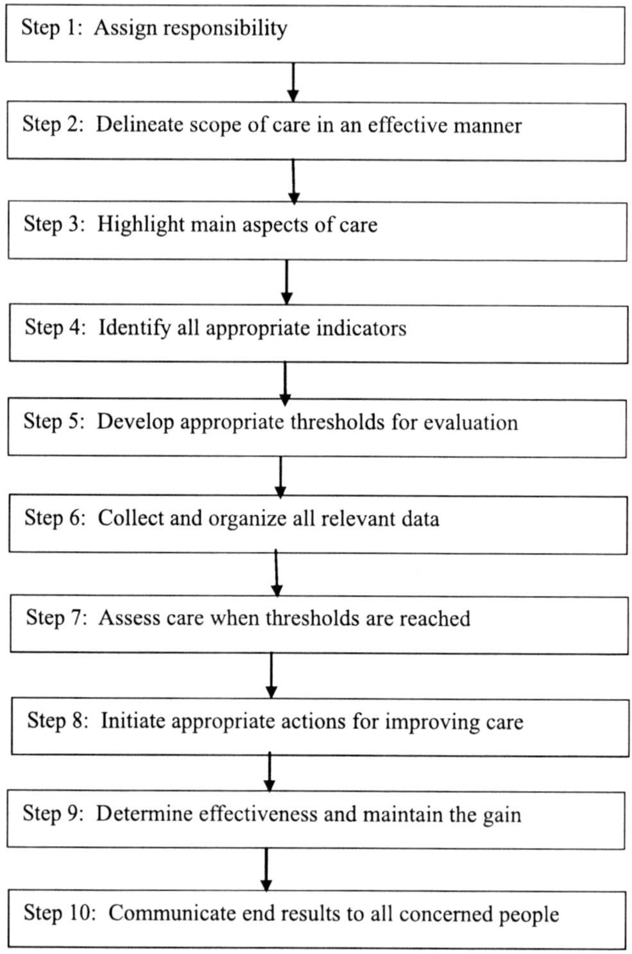

Figure 8.3. Steps for improving quality in health care

8.5 Quality Tools for Use in Health Care

There are many methods that can be used to improve quality in health care. Most of these methods are listed in Table 8.3 [5, 12]. The first five of these methods are described below (information on others can be found in Chapters 3 and 11, or in Refs. [5, 12, 13]).

Table 8.3. Methods for improving quality in health care

No.	Method
1	Brainstorming
2	Cost-benefit analysis
3	Multivoting
4	Force field analysis
5	Check sheets
6	Cause and effect diagram
7	Scatter diagram
8	Pareto chart
9	Histogram
10	Control chart
11	Process flowchart
12	Affinity diagram
13	Prioritization matrix
14	Proposed options matrix

8.5.1 Group Brainstorming

The objective of brainstorming in health care quality is to generate ideas, options or identify problems, concerns. It is often referred to as a form of divergent thinking because the basic purpose is to enlarge the number of ideas being considered. Thus, brainstorming may simply be described as a group decision-making approach designed to generate many creative ideas by following an interactive process. The team concerned with health care quality can make use of brainstorming to get its ideas organized into a quality method such as a cause and effect diagram or a process flow diagram.

Past experiences indicate that questions such as listed below can be quite useful to start a brainstorming session concerned with health care quality [12].

- What are the major obstacles to improving quality?
- What are the health care organization's three most pressing unsolved quality problems?
- What type of action plan is required to overcome these problems?
- What are the most pressing areas that require such action plan?

Some of the useful guidelines for conducting effective brainstorming sessions are shown in Figure 8.4 [14, 15].

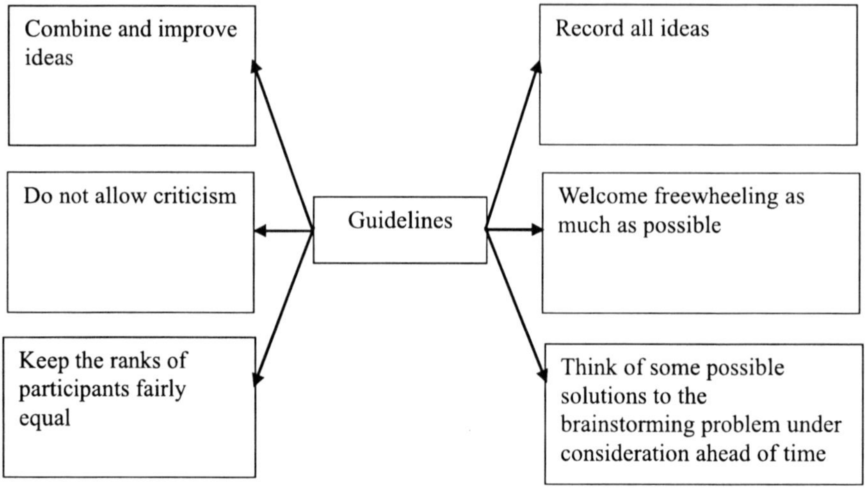

Figure 8.4. Useful guidelines for conducting effective brainstorming sessions

8.5.2 Cost–Benefit Analysis

Cost–benefit analysis may simply be described as a weighing-scale approach to decision-making, where all plusses (*i. e.*, cash flows and other intangible benefits) are grouped and put on one side of the balance and all the minuses (*i. e.*, costs and drawbacks) are grouped and put on the other. At the end the heavier side wins.

The main purpose of the application of the cost–benefit analysis method in the health care quality area is that the quality team members consider the total impact of their recommended actions. Additional information on this method is available in Refs. [2, 16–17].

8.5.3 Multivoting

This is useful method for reducing a large number of ideas to a manageable few judged important by the participating individuals. Usually by following this approach, the number of ideas is reduced to three to five [2]. Another thing that can be said about multivoting is that it is a form of convergent thinking because the objective is to reduce the number of ideas being considered. Needless to say, multivoting is considered to be a useful tool for application in the health care quality area, and additional information on the method is available in Ref. [18].

8.5.4 Force Field Analysis

This method was developed by Kurt Lewin for identifying the forces that are related to a certain issue under consideration [13, 19]. The method is also known as barriers and aids analysis [2]. In this approach, the issue/problem statement is written at the top of a sheet and two columns are created below it for writing negative forces on one side and the positive on the other.

Subsequently, these forces are ranked and appropriate ways and means to mitigate the negative forces and accentuate the positive forces are explored. Additional, information on the method is available in Refs. [2, 13, 19].

8.5.5 Check Sheets

Check sheets are basically used for collecting data on occurrence frequency of specified events. A check sheet, for example, can be utilized in determining the occurrence frequency of, say, four to six problems highlighted during multivoting [2]. In the quality areas, check sheets are usually used in a quality improvement process for collecting frequency-related data later displayed in a Pareto diagram.

Although there is no standard design of check sheets, the basic idea is to document all types of important information relative to nonconformities and nonconforming items, so that the sheets can facilitate improvement in the process. Additional information on check sheets is available in Refs. [20–22].

8.6 Implementation of Six Sigma Methodology in Hospitals and Its Potential Advantages and Implementation Barriers

The history of Six Sigma as a measurement standard may be traced back to Carl Frederick Gauss (1777–1855), the father of the concept of the normal curve. In the 1980s Motorola explored this standard and created the methodology and necessary cultural change associated with it.

Six Sigma may simply be described as a methodology implementation directed at a measurement-based strategy that develops process improvements and varied cost reductions throughout an organizational set up. In many organizations, Six Sigma simply means a measure of quality that strives for near perfection.

Over the past few years, a number of health care organizations have also started to apply the Six Sigma methodology into their operations. A total of nine steps, as shown in Figure 8.5, are involved in the implementation of define, measure, analyze, improve, control (DMAIC) Six Sigma methodology in an industrial organization [23]. These steps can be tailored accordingly for the implementation of the methodology in hospitals.

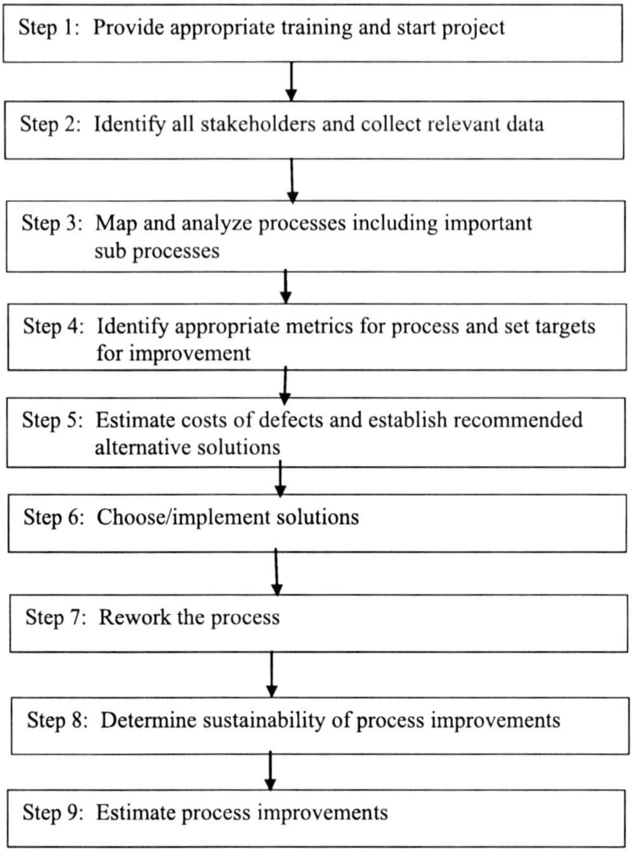

Figure 8.5. Steps involved in the implementation of DMAIC Six Sigma methodology

Some of the important potential advantages of implementation of Six Sigma methodology in hospitals are as follows [23]:

- Measurement of essential health care performance requirements on the basis of commonly used standards.
- Establishment of shared accountability with respect to continuous quality improvement.
- The implementation of the methodology with emphasis on improving customers' lives, could result in the involvement of more health care professionals and support personnel in the quality improvement effort.
- Better job satisfaction of health care employees.

There are many potential barriers to the implementation of Six Sigma programs in hospitals. Some of these are shown in Figure 8.6 [23].

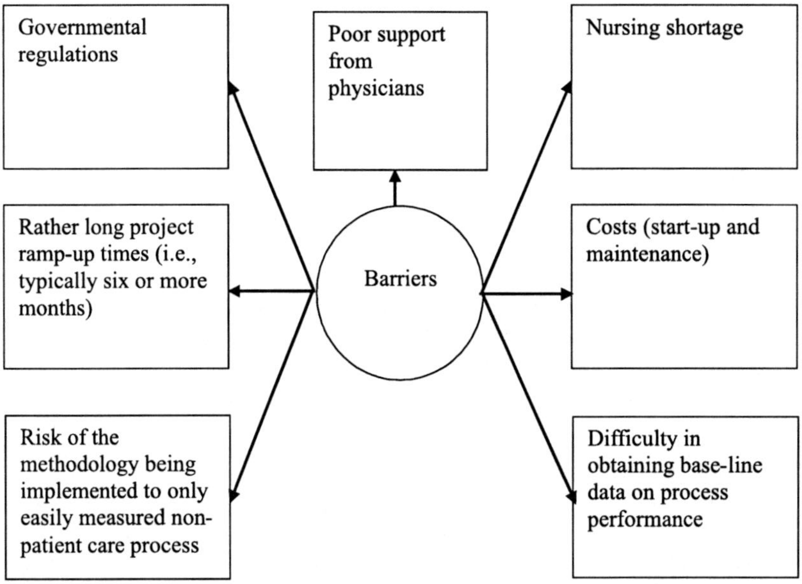

Figure 8.6. Potential barriers to the implementation of Six Sigma methodology in hospitals

8.7 Problems

1. Write a short essay on the historical developments in quality in health care.
2. Define the following three terms:
   - Quality of care
   - Health care
   - Clinical audit
3. What are the main reasons for rising health care costs?
4. Compare traditional quality assurance and total quality management with respect to health care.
5. Discuss health care-related quality goals.
6. Discuss physician reactions to total quality management.
7. List at least ten quality tools useful for application in the health care sector.
8. Discuss the implementation of Six Sigma methodology in hospitals and its benefits.
9. What are the ten useful steps for improving quality in health care.
10. Discuss the following three methods considered useful for improving quality in health care:
    - Force field analysis
    - Multivoting
    - Group brainstorming

References

1 Graham, N.O., Quality Trends in Health Care, in Quality in Health Care, edited by N.O. Graham, Aspen Publishers, Gaithersburg, Maryland, 1995, pp. 3–14.
2 Gaucher, E.J., Coffey, R.J., Total Quality in Health Care: from Theory to Practice, Jossey-Bass Publishers, San Francisco, 1993.
3 Codman, E.A., The Product of the Hospital, Surgical Gynaecology and Obstetrics, Vol. 28, 1914, pp. 491–496.
4 Glossary of Terms Commonly Used in Health Care, Prepared by the Academy Health, Suite 701-L, 1801 K St. NW, Washington, D.C., 2004.
5 Graham, N.O., Editor, Quality in Health Care: Theory, Application, and Evolution, Aspen Publishers, Gaithersburg, Maryland, 1995.
6 Marszalek-Gaucher, E., Coffey, R.J., Transforming Health Care Organizations: How to Achieve and Sustain Organizational Excellence, John Wiley and Sons, New York, 1990.
7 Berwick, D.M., Peer Review and Quality Management: Are They Compatible?, Quality Review Bulletin, Vol. 16, 1990, pp. 246–251.
8 Fainter, J., Quality Assurance Not Quality Improvement, Journal of Quality Assurance, January/February, 1991, pp. 8, 9, and 36.
9 Coltin, K.L., Aronow, D.B., Quality Assurance and Quality Improvement in the Information Age, in Quality in Health Care: Theory, Application, and Evolution, edited by N.O. Graham, Aspen Publishers, Gaithersburg, Maryland, 1995.
10 Andrews, S.L., QA versus QI: The Changing Role of Quality in Health Care, January/February, 1991, pp. 14, 15, and 38.
11 Laffel, G., Blumenthal, D., The Case for Using Industrial Quality Management Science in Health Care Organization, Journal of the American Medical Association, Vol. 262, 1989, pp. 2869–2873.
12 Stamatis, D.H., Total Quality Management in Health Care, Irwin Professional Publishing, Chicago, 1996.
13 Dhillon, B.S., Creativity for Engineers, World Scientific Publishing, River Edge, New Jersey, 2006.
14 Osborn, A.F., Applied Imagination, Charles Scribner's Sons, New York, 1963.
15 Dhillon, B.S, Engineering and Technology Management Tools and Applications, Artech House, Inc., Boston, 2002.
16 Boardman, A.E., Cost-Benefit Analysis: Concepts and Practice, Prentice Hall, Upper Saddle River, New Jersey, 2006.
17 Levin, H.M., McEwan, P.J., Cost-Effectiveness Analysis: Methods and Applications, Sage Publications, Thousand Oaks, California, 2001.
18 Tague, N.R., The Quality Toolbox, ASQ Quality Press, Milwaukee, Wisconsin, 2005.
19 Jay, R., The Ultimate Book of Business Creativity: 50 Great Thinking Tools, for Transforming Your Business, Capstone Publishing Limited, Oxford, U.K., 2000.
20 Ishikawa, K., Guide to Quality Control, Asian Productivity Organization, Tokyo, 1976.
21 Montgomery, D.C., Introduction to Statistical Control, John Wiley and Sons, New York, 1996.
22 Leitnaker, M.G., Sanders, R.D., Hild, C., The Power of Statistical Thinking: Improving Industrial Processes, Addison-Wesley, Reading, Massachusetts, 1996.
23 Frings, G.W., Graut, L., Who Moved My Sigma – Effective Implementation of the Six Sigma Methodology to Hospitals, Quality and Reliability Engineering International, Vol. 21, 2005, pp. 311–328.

9

Software Quality

9.1 Introduction

Today computers are widely used for applications ranging from-day-to day personal use to control of space systems. As the computers are made up of both hardware and software elements, the proportion of the total computer cost spent on software has changed quite dramatically over the years. For example, in 1955 the software component (*i. e.*, including software maintenance) accounted for 20% of the total computer cost and 30 years later, in 1985, this percentage has increased to 90% [1]. Needless to say, the introduction of computers into products in the late 1970s has led to the software quality assurance for all types of software [2].

Furthermore, it can be added that no product is of greater quality than the quality of its elements, and if one of the elements is a computer, then the quality of software or program controlling that computer will certainly affect the quality of the product. The prime objective of a quality assurance program is to assure that the end software products are of good quality, through properly planned and systematic activities or actions to achieve, maintain, and determine that quality [3–4]. This chapter presents various different important aspects of software quality.

9.2 Software Quality Terms and Definitions

There are many terms and definitions used in the software quality area. Some of the commonly used terms and definitions are as follows [2, 5–7]:

- **Software quality.** This is the fitness for use of the software item/product.
- **Software quality control.** This is the independent evaluation of the capability of the software process to produce a usable software product/item.

- **Software quality assurance.** This is the set of systematic activities or actions providing evidence of software process's capability to produce a software product/item that is fit to use.
- **Software quality testing.** This is a systematic series of evaluation actions or activities carried out to validate that the software fully satisfies performance and technical requirements.
- **Software reliability.** This is the ability of the software to carry out its specified function under stated conditions for a given period of time.
- **Software maintenance.** This is the process of modifying a software system or element after delivery, to rectify faults, enhance performance or other appropriate attributes, or adapt to a changed environment.
- **Verification and validation.** This is the systematic process of analyzing, evaluating, and testing system and software code and documentation for ensuring maximum possible reliability, quality, and satisfaction of system needs and goals.
- **Software process management.** This is the effective utilization of available resources both to produce properly engineered products/items and to enhance the software engineering capability of the organization.
- **Software process improvement.** This is a deliberate, planned methodology following standardized documentation practices for capturing on paper (and in practice) the approaches, activities, practices, and transformations that individuals use for developing and maintaining software and the associated products.
- **Software.** This is computer programs, procedures, and possibly associated data and documentation pertaining to the operation of a computer.

9.3 Software Quality Factors and Their Subfactors

The large variety of issues concerning various attributes of software and its use and maintenance, as outlined in software requirement documentation, may be categorized into content groups known as quality factors. Over the years, many models of software quality factors and their classification in factor categories have been proposed by various authors [8]. One of these models classifies all software requirements into 11 software quality factors grouped under 3 categories as shown in Figure 9.1 [8]. These categories are product operation factors (Category I), product revision factors (Category II), and product transition factors (Category III).

The product operation factors are concerned with requirements that directly affect the daily operation of the software. Five specific factors that belong to this category are correctness, usability, integrity, reliability, and efficiency. The product revision factors are concerned with requirements that affect all software maintenance activities: adaptive maintenance (*i. e.*, adapting the current software to additional customers and circumstances without making many changes to the software), perfective maintenance (*i. e.*, enhancing and improving the current software with respect to locally limited issues), and corrective maintenance (*i. e.*, correcting software faults and failures). Three specific factors that belong to this category are testability, maintainability, and flexibility.

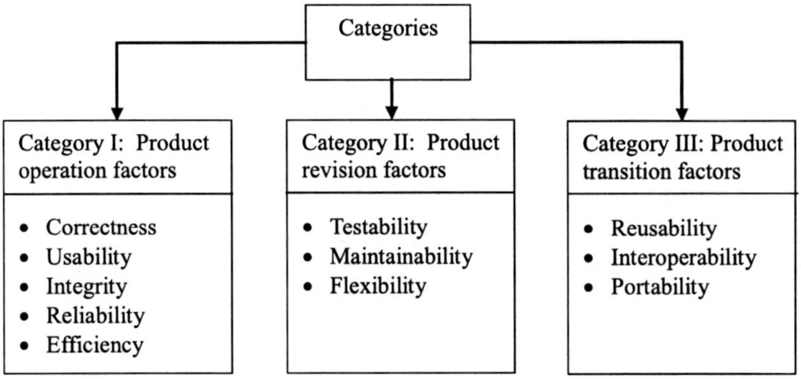

Figure 9.1. Three categories of the software quality factors

The product transition factors are concerned with the adaptation of software to other environments as well as its interaction with other software systems. Three specific factors that belong to this category are reusability, interoperability, and portability.

Each of the above specific software quality factors belonging to Categories I, II, and III are discussed below [9, 10].

- **Correctness.** Correctness requirements are outlined in a list of required outputs of the software system. The subfactors of the correctness are completeness, availability (response time), accuracy, up-to-dateness, compliance (consistency), and coding and documentation guidelines.
- **Usability.** Usability requirements are concerned with the scope of staff resources required for training a new employee as well as to operate the software system. Two subfactors of the usability are operability and training.
- **Integrity.** Integrity requirements are concerned with the software system security, *i. e.*, requirements for preventing access to unauthorized individuals, to distinguish between the majority of individuals permitted to view the information ("read permit") and a limited number of individuals who will be permitted to add and change data ("write permit"), etc. Two subfactors of the integrity are access control and access audit.
- **Reliability.** Reliability requirements are concerned with failures to provide an appropriate level of service. Furthermore, they determine the maximum permitted failure rate of the software system and can refer to the total system or one or more of its separate functions. Four subfactors of the reliability are system reliability, hardware failure recovery, application reliability, and computational failure recovery.

- **Efficiency.** Efficiency requirements are concerned with the hardware resources required for carrying out the entire functions of the software system, in conformance to all other requirements. Four subfactors of the efficiency are efficiency of processing, efficiency of storage, efficiency of communication, and efficiency of power usage (for portable units).
- **Testability.** Testability requirements are concerned with the testing of an information system as well as with its specified operation. Three subfactors of the testability are traceability, user testability, and failure maintenance testability.
- **Maintainability.** Maintainability requirements are concerned with determining the efforts that will be required by all potential users and maintenance people for identifying the reasons for the occurrence of software failures, to rectify the failures, and to verify the success of the rectifications or corrections. Six subfactors of the maintainability are modularity, simplicity, compliance (consistency), document accessibility, coding and documentation guidelines, and self-descriptiveness.
- **Flexibility.** Flexibility requirements are concerned with the capabilities and efforts needed to support adaptive maintenance activities. Four subfactors of the flexibility are simplicity, modularity, generality, and self-descriptiveness.
- **Reusability.** Reusability requirements are concerned with the use of software modules, originally designed for one particular project, in a new software project being developed. Seven subfactors of the reusability are simplicity, generality, modularity, document accessibility, self-descriptiveness, application independence, and software system independence.
- **Interoperability.** Interoperability requirements are concerned with creating interfaces with other software systems or with other equipment/product firmware. Four subfactors of the interoperability are modularity, commonality, system compatibility, and software system independence.
- **Portability.** Portability requirements are concerned with the adaptation of a software system in question to other environments composed of different operating systems, different hardware, etc. Three subfactors of the portability are modularity, self descriptive, and software system independence.

9.4 Useful Quality Tools for Use During the Software Development Process

There are many quality tools that can be used, to improve software quality, during the software development process. Seven of these tools are listed in Table 9.1 [11]. Some of these are briefly described below (detailed information of these or other quality tools can be found in Chapters 3, 8, 11, or in Refs [12–14]).

- **Run charts.** Run charts are often used for software project management and serve as real-time statements of quality as well as work load. An example of the application of run charts is the monitoring of weekly arrival of software defects and defect backlog during the formal machine testing phases. Another example of the run chart application is tracking the percentage of software fixes that exceed the fix response time criteria, in order to ensure timely deliveries of fixes to customers. Needless to say, during the software development process, often run charts are compared to the projection models and historical data so that all the associated interpretations can be placed into appropriate perspectives. Additional information on run charts with respect to their application during the software development process is available in Ref. [11].

- **Pareto diagram.** Pareto diagrams are probably the most applicable tool in the software quality area because past experiences indicate that software defects or defect density never follow a uniform distribution. Pareto diagrams are an effective tool to identify focus areas that cause most of the problems in a software project under consideration. For example, Motorola successfully used the Pareto diagram to identify main sources of software requirement changes that enabled in-process corrective measures to be taken [15]. Another example is that Hewlett-Packard through Pareto analysis was able to achieve significant software quality improvements [16]. Additional information on Pareto diagrams with respect to their application during the software development process is available in Ref. [11].
- **Checklists.** Checklists play a significant role in the software development process because they are useful to software developers/programmers to ensure that all tasks are complete, and for each of these tasks the important factors or quality characteristics are taken into consideration. The use of checklists is quite pervasive. Checklists, used daily by the software development people, are developed and revised on the basis of accumulated experience.

Checklists are frequently an element of the process documents, and past experiences indicate that their daily application is quite useful to keep the software development processes alive. Additional information on check sheets with respect to their application during the software development process, is available in Ref. [11].

Table 9.1. Quality tools for use during the software development process

No.	Quality tools
1	Scatter diagram
2	Run charts
3	Control chart
4	Checklist
5	Histogram
6	Cause and effect diagram
7	Pareto diagram

9.5 A Manager's Guide to Total Quality Software Design

In order to have good quality end software products, it is important to take proper quality-related measures during the software development life cycle (SDLC). A SDLC includes five stages as shown in Figure 9.2 [17]. Each of these stages with respect to assuring quality is discussed below, separately.

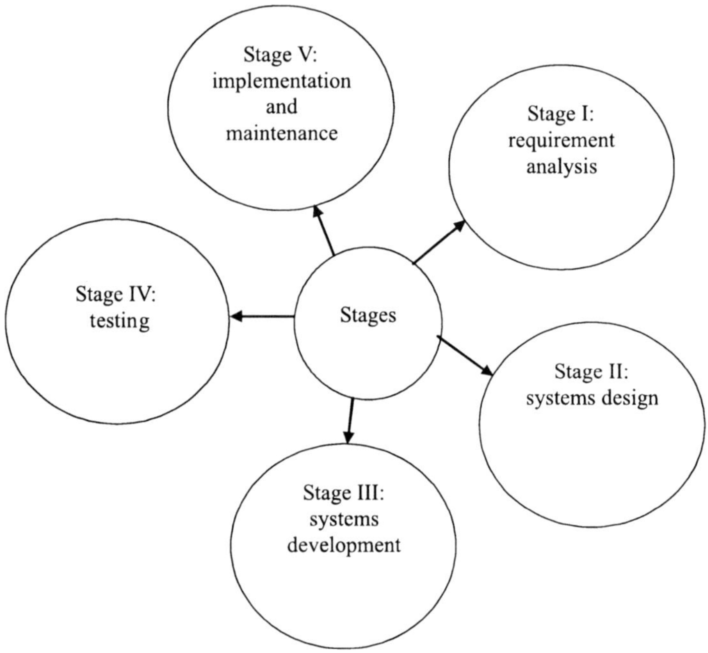

Figure 9.2. Software development life cycle (SDLC) stages

9.5.1 Stage I: Requirements Analysis

Over the years, it has been estimated that around 60–80% of system development failures are the result of poor understanding of user requirements [18]. In this regard, usually major software vendors make use of the quality function development (QFD) during the software development process. Software quality function deployment (SQFD) is a useful tool for focusing on improving the quality of the software development process by implementing quality improvement approaches to the SDLC requirements solicitation phase. More specifically, SQFD is a front-end requirements collection approach that quantifiably solicits and defines critical customer requirements. Thus it is a quite useful tool to solve the problem of poor systems specification during SDLC. Some of the main advantages of SQFD are establishing better communications among departments and with customers, quantifying qualitative customer requirements, fostering better attention to customers' requirements, and reaching features consensus faster [17].

9.5.2 Stage II: Systems Design

This is the most critical stage of quality software development because a defect in design is hundreds of times more costly to rectify than a defect during the production stage. More specifically, it basically means that every dollar spent to increase design quality has at least a hundred-fold payoff during the implementation and operation stages [19]. Concurrent engineering is a widely used method to change systems design and also it is a useful method of implementing total quality management [17]. Additional information on concurrent engineering is available in Refs. [20, 21].

9.5.3 Stage III: Systems Development

Software total quality management (TQM) calls for the integration of quality into the total software development process. After the establishment of a quality process into the first two stages of software development cycle, the task of coding becomes much easier [17]. Nonetheless, for document inspections, the method of design and code inspections can be used [22]. Furthermore, for tracking the metrics of the effectiveness of code inspections, control charts can be used.

9.5.4 Stage IV: Testing

Testing activities should be properly planned and managed from the start of software development, in addition to designing them properly at each stage of the software development life cycle [23]. Nonetheless, a TQM-based software testing process must have a clear set of testing objectives. A six step metric-driven method can fit with such testing objectives. Its steps are establish structured test objectives, select appropriate functional methods to derive test case suites, run functional tests and assess the degree of structured coverage achieved, extend the test suites until the achievement of the desired coverage, calculate the test scores, and validate testing by recording errors not discovered during testing [17].

9.5.5 Stage V: Implementation and Maintenance

Most of the software maintenance activities are reactive. More specifically, programmers frequently zero in on the immediate problem, fix it, and wait until the occurrence of the next problem [17, 24]. As statistical process-control (SPC) can be used to monitor the quality of software system maintenance, a TQM-based system must adapt to the SPC process to assure maintenance quality. Additional information concerning quality during software maintenance is available in Refs. [17, 25].

9.6 Software Quality Metrics

There is a large number of metrics that can be used to improve or assure software quality. Two main objectives of software quality metrics are to highlight conditions that need or enable development or maintenance process improvement in the form of corrective or preventive measures initiated within the organization and to facilitate an appropriate level of management control including planning and executing of proper management interventions.

For their successful applicability, it is essential that these metrics must satisfy requirements such as comprehensive (*i. e.*, applicable to a wide variety of implementations and situations), reliable (*i. e.*, generate similar results when applied under similar environments), valid (*i. e.*, successfully measure the required attribute), relevant (*i. e.*, related to an attribute of substantial importance), mutually exclusive (*i. e.*, do not measure attributes measured by other metrics), easy and simple (*i. e.*, the implementation of the metrics data collection is simple and straight forward and is carried out with minimal resources), do not require independent data collection, and immune to biased interventions by interested parties [9]. Some of the software quality metrics are presented below [9, 26].

9.6.1 Metric I

Metric I is one of the error density metrics and is expressed by

$$CE_d = \frac{TN_{ce}}{LC} \tag{9.1}$$

where

CE_d is the code error density.

LC is the thousands of lines of code.

TN_{ce} is the total number of code errors detected in the software code through inspections and testing. Data required for this measure are obtained from code inspection and testing reports.

9.6.2 Metric II

Metric II is one of the error severity metrics and is expressed by

$$CE_{as} = \frac{WCED}{TN_{ce}} \tag{9.2}$$

where

CE_{as} is the average severity of code errors.

$WCED$ is the weighted code errors detected. Data required for this measure are also obtained from code inspection and testing reports.

9.6.3 Metric III

Metric III is one of the error removal effectiveness metrics and is defined as follows:

$$DE_{re} = \frac{NDCE}{NDCE + TNSFD} \tag{9.3}$$

where

DE_{re} is the development error removal effectiveness.

$TNSFD$ is the total number of software failures detected during a one year period of maintenance service.

$NDCE$ is the number of design and code errors detected in the software development process. Usually data for this measure are obtained from design and code reviews and testing reports.

9.6.4 Metric IV

Metric IV is one of the software process timetable metrics and is expressed by

$$TT_{of} = \frac{TNMC}{TNM} \tag{9.4}$$

where

TT_{of} is the time table observance factor.

TNM is the total number of milestones.

$TNMC$ is the total number of milestones completed on time.

9.6.5 Metric V

Metric V is one of the software process productivity metrics and is defined by

$$SDP = \frac{HIDSS_w}{LC} \tag{9.5}$$

where

SDP is the software development productivity.

$HIDSS_w$ is the total number of working hours invested in the development of the software system.

9.6.6 Metric VI

Metric VI is one of the help desk service (HDS) calls density metrics and is expressed as follows:

$$HDSCD = \frac{TNHDC}{LMSC} \qquad (9.6)$$

where
HDSCD is the HDS calls density.
LMSC is the thousands of lines of maintained software code.
TNHDC is the total number of HDS calls during one year period of service.

9.6.7 Metric VII

Metric VII is concerned with measuring the success of the HDS and is defined by

$$HDSS = \frac{NHDSC}{TNHDC} \qquad (9.7)$$

where
HDSS is the HDS success factor.
NHDSC is the total number of HDS calls completed on time during one year period of service.

9.6.8 Metric VIII

Metric VIII is concerned with measuring the average severity of the HDS calls and is expressed by

$$ASHDSC = \frac{NWHDSC}{TNHDC} \qquad (9.8)$$

where
ASHDSC is the average severity of HDS calls.
NWHDSC is the total number of weighted HDS calls received during one year period of service.

9.6.9 Metric IX

Metric IX is one of the HDS productivity metrics and is defined as follows:

$$HDSP = \frac{HDSWHS}{LMSC} \qquad (9.9)$$

where
HDSP is the HDS productivity factor.
HDSWHS is the number of annual working hours invested in help desk servicing of the software system.

9.6.10 Metric X

Metric X is concerned with measuring the software corrective maintenance effectiveness and is expressed by

$$CME = \frac{CMHS}{TNSF} \tag{9.10}$$

where
CME is the corrective maintenance effectiveness.
CMHS is the number of annual working hours invested in the corrective maintenance of the software system.
TNSF is the total number of software failures detected during one year period of maintenance service.

9.7 Software Quality Cost

Software quality cost can be classified as shown in Figure 9.3 [9]. The figure shows two main classifications (*i. e.*, cost of controlling failures and cost of the failure of control) and four subclassifications (*i. e.*, prevention costs, appraisal costs, internal failure costs, and external failure costs.)

The cost of controlling failures is associated with activities performed to detect and prevent software errors, in order to reduce them to an acceptable level. Two subcategories of the cost of controlling failures are prevention costs and appraisal costs. Prevention costs are associated with activities such as developing a software quality infrastructure, improving and updating that infrastructure, and carrying out the regular activities required for its operation. Appraisal costs are concerned with activities pertaining to the detection of software errors in specific software systems/projects. Typical components of appraisal costs are the cost of reviews, cost of software testing, and cost of assuring quality of external participants (*e. g.*, subcontractors).

The cost of the failure of control is concerned with the cost of failures that occurred because of failure to detect and prevent software errors. Two subcategories of the cost of the failure of control are internal failure costs and external failure costs. Internal failure costs are associated with correcting errors found through design reviews, software tests, and acceptance tests, prior to the installation of the software at customer sites. Similarly, the external failure costs are associated with correcting failures detected by customers/maintenance teams after the installation of the software system at customer sites.

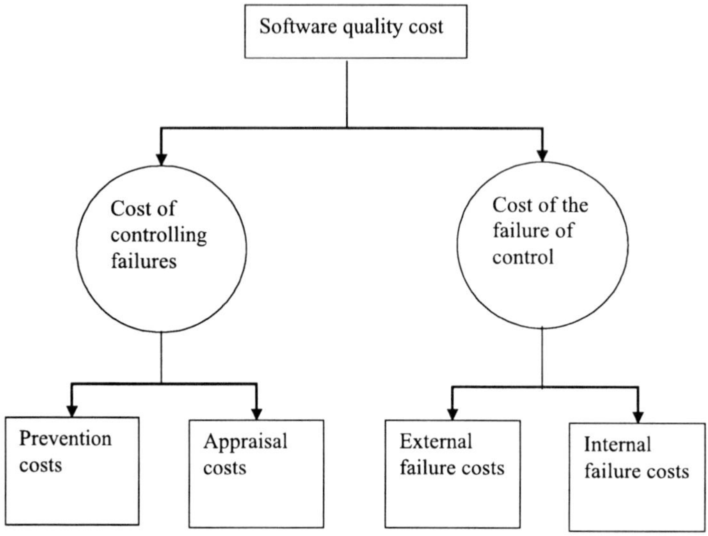

Figure 9.3. Classifications and subclassifications of software quality cost

9.8 Problems

1. Write an essay on software quality.
2. Define the following three terms:
   - Software quality control
   - Software quality
   - Software quality assurance
3. What are the software quality factors? List at least nine of them.
4. Describe run charts.
5. List at least four quality tools that can be used during the software development process.
6. What are the main stages of the software development life cycle?
7. What is a software metric?
8. Define at least three software quality metrics.
9. Define software quality cost.
10. What are the four categories of the software quality cost?

References

1 Keene, S.J., Software Reliability Concepts, Annual Reliability and Maintainability Symposium Tutorial Notes, 1992, pp. 1–21.
2 Dunn, R., Ullman, R., Quality Assurance for Computer Software, McGraw-Hill Book Company, New York, 1982.
3 Mendis, K.S., A Software Quality Assurance Program for the 80s, Proceedings of the Annual Conference of the American Society for quality Control, 1980, pp. 379–388.
4 Dhillon, B.S., Reliability in Computer System Design, Ablex Publishing Corporation, Norwood, New Jersey, 1987.
5 Schulmeyer, G.G., Software Quality Assurance: Coming to Terms, in Handbooks of Software Quality Assurance, edited by G.G., Schulmeyer and J.I. McManus, Prentice Hall, Inc., Upper Saddle River, New Jersey, 1999, pp. 1–27.
6 IEEE-STD-610.12-1990, IEEE Standard Glossary of Software Engineering Terminology, Institute of Electrical and Electronics Engineers (IEEE), New York, 1991.
7 Ralston, A., Reilly, E.D., Editors, Encyclopaedia of Computer Science, Van Nostrand Reinhold Company, New York, 1993.
8 McCall, J., Richards, P., Walters, G., Factors in Software Quality, NTIS Report No. AD-A049-014, 015, 055, November 1977. Available from the National Technical Information Service (NTIS), Springfield, Virginia, USA.
9 Galin, D., Software Quality Assurance, Person Education Ltd., Harlow, Essex, U.K., 2004.
10 Evans, M.W, Marciniak, J.J., Software Quality Assurance and Management, John Wiley and Sons, New York, 1987.
11 Kan, S.H., Metrics and Models in Software Quality Engineering, Addison-Wesley Publishing Company, Reading, Massachusetts, 1995.
12 Ishikawa, K., Guide to Quality Control, Asian Productivity Organization, Tokyo, 1976.
13 Mears, P., Quality Improvement tools and Techniques, McGraw-Hill Book Company, New York, 1995.
14 Kanji, G.K., Asher, M., 100 Methods for Total Quality Management, Sage Publications Ltd., London, 1996.
15 Daskalantonakis, M.K., A Practical View of Software Measurement and Implementation Experiences within Motorola, IEEE Transactions on Software Engineering, Vol. SE-18, 1992, pp. 998–1010.
16 Grady, R.B., Caswell, D.L., Software Metrics: Establishing a Company-Wide Program, Prentice Hall, Inc., Englewood Cliffs, New Jersey, 1986.
17 Gong, B., Yen, D.C., Chou, D.C., A Manager's Guide to Total Quality Software Design, Industrial Management and Data Systems, March 1998, pp. 100–107.
18 Gupta, Y.P., Directions of Structured Approaches in System Development, Industrial Management and Data Analysis, July/August 1988, pp. 11–18.
19 Zahedi, F., Quality Information Systems, Boyd and Fraser, Inc., Danvers, Massachusetts, 1995.
20 Rosenblatt, A., Watson, G.F., Concurrent Engineering, IEEE Spectrum, July 1991, pp. 22–23.
21 Salomone, T.A., Concurrent Engineering, Marcel Dekker, Inc., New York, 1995.

22 Fagan, M.E., Advances in Software Inspection, IEEE Transactions on Software Engineering, Vol. 12, No. 7, 1986, pp. 744–751.
23 Graham, D.R., Testing and Quality Assurance – the Future, Information and Software Technology, Vol. 34, No. 10, 1992, pp. 694–697.
24 Osborne, W.M., All About Software Maintenance: 50 Questions and Answers, Journal of Information Systems Management, Vol. 5, No. 3, 1988, pp. 36–43.
25 Schulmeyer, G.G., McManus, J.I., Total Quality Management for Software, Van Nostrand Reinhold, New York, 1992.
26 Schulmeyer, G.G., Software Quality Assurance Metrics, in the Handbook of Quality Assurance, edited by G.G. Schulmeyer and J.I. McManus, Prentice Hall, Inc., Upper Saddle River, New Jersey, 1999, pp. 403–443.

10

Quality Control in the Textile Industry

10.1 Introduction

Each year billions of dollars are spent to produce various types of textiles, and the world production of fibres was predicted to be around 50 million tons for the year 2,000. The United States has only 4.3 percent of the world's population, but it consumes almost 20 percent of the world's textiles [1,2].

The history of quality control in the textile industry may be traced back to Zhou Dynasty (11th to 8th centuries B.C.) in China. For example, one dynasty decree stated that "Cottons and silks of which the quality and size are not up to the standards are not allowed to be sold on the market" [2,3]. In the modern context, the first application of statistical quality control concepts appeared to be in yarn-manufacturing products during the late 1940s and 1950s [2]. In 1981, one of the largest textile companies in the world, Milliken & Company, launched its total quality management efforts specifically directed to make a commitment to customer satisfaction pervading all company levels and locations. By 1989, it was ahead of its competition with respect to all measures of customer satisfaction in the United States and won the Baldridge Quality Award [4].

Currently, there are around 30,000 textile-related companies in the United States, and many of them have implemented quality management initiatives for reducing costs and improving both products and customer satisfaction. This chapter presents various important aspects of quality control in the textile industry.

10.2 Quality-related Issues in Textiles and Quality Problems Experienced in Apparel

Some of the quality-related issues directly or indirectly associated with textiles are as follows [5–6]:

- Poor understanding of the customer needs and satisfaction.
- Inadequate training of operators in their jobs and in quality issues.
- Quality is not a pressing issue until it becomes a problem.
- Often quality enters downstream, more specifically, at final assembly, rather than in the design and development stages (*i. e.*, early stages).
- Quality-related problems are observed with vendors.
- Quality costs are considered to be too high.
- Management appears to sacrifice quality when costs and scheduling conflict.

The main quality problems, and their corresponding percentages in parentheses, experienced in apparel are shown in Figure 10.1 [7]. These are material failure, construction/stitching failure, customer misuse, and faulty trimmings. Material failure consists of loose dye during exposure to items such as washing, sea water, rubbing, ironing, perspiration, water, dry cleaning, light, and chlorinated water; dimensional instability due to shrinkage and stretching; and poor wear and appearance due to factors such as slippage, pilling, abrasion, and snagging. Construction failure consists of faulty seams due to factors such as wrong machine settings, poor quality of design, incorrect machinery relative to fabric, and weak sewing thread and faulty interlining due to delamination. Customer misuse includes unfair wear and tear and wrong washing methods.

Faulty trimmings consist of broken fasteners, broken zips, buttons incorrectly stitched, and button dye.

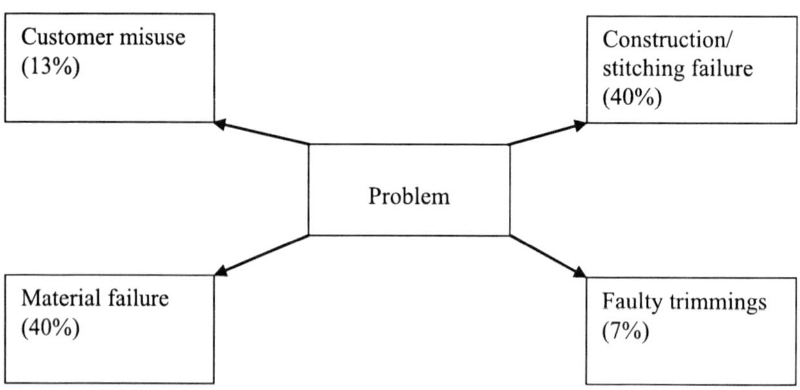

Figure 10.1. Main quality problems experienced in apparel

10.3 Fibres and Yarns

The basic raw materials of the textile industry are fibres. Normally, they are transformed into yarn and then into fabric. There are two main categories of fibres: naturally occurring and man made [8]. The natural fibres may be of vegetables (*e. g.*, cotton and linen), animal (*e. g.*, hair and wool), or mineral (*e. g.*, asbestos) origin. The man-made fibres can be grouped as synthetic polymers (*e. g.*, polypropylene, polyester, etc.), natural polymers (*e. g.*, acetate, viscose rayon, etc.), and others (*e. g.*, glass, metal, carbon, etc.).

Besides the above two broad categories of fibres, additional classification is made by measuring fibre density, checking extensibility, observing any reaction to staining, testing for the presence of certain elements such as chlorine and nitrogen, a drying twist test, differential thermal analysis, a refractive index test for glass fibres, treating with a series of solvents either at room temperature or the boiling point until one is found in which the fibres dissolve, etc. [8].

Although each category of fibre may be used individually or in blends, nowadays it is a common practice to blend natural with man-made fibres for achieving an optimum combination of physical properties and cost/price.

The physical properties of fabrics and yarns are subject to various factors including the fibre properties. A careful analysis of fibre properties along with experience, can give a broad idea about the likelihood of the end result when the fibres are spun into yarn.

A yarn may simply be described as an assembly of fibres and/or filaments normally bound together by twist. The basic specification of a yarn includes at least the materials, the twist, and the count. Nonetheless, some of the important items pertaining to yarn are as follows [8]:

- **Count.** Yarn count may be described directly as mass per unit length or indirectly as length per unit mass.
- **Diameter.** Diameter is a measure of yarn covering power (*i. e.*, the extent to which cloth area is covered by single set of threads) when in the fabric and is measured by throwing yarn shadow (silhouette) on to a graduated glass scale.
- **Twist.** This may simply be described as the number of turns per unit length of yarn. It causes frictional trappings between the fibres and hence imparts appropriate strength to the yarn. The twist factor is a measure of twist hardness and it relates twist to the linear density. It is expressed by

$$TF = (twist)(linear\ density)^{\frac{1}{2}} \tag{10.1}$$

  where
  TF is the twist factor.

- **Friction.** The tension of yarn in processing basically depends on friction as it passes around machinery parts or yarn guides. The coefficient of friction of a yarn as it runs around a test object is measured by the yarn friction tester.

- **Crimp of yarn in fabric.** Precise and accurate measurement of yarn length is absolutely important in estimating crimp (take up, regain, shrinkage, etc.) in woven fabrics, in measuring the count of short lengths of yarn, and in calculating course and loop-length in knitted fabrics. The following formula is used to calculate crimp in percentage [8]:

$$C = \frac{(SL - LIF)(100)}{LIF} \tag{10.2}$$

where
C is the crimp expressed in percentage.
SL is the straightened length.
LIF is the length in fabric.

10.4 Textile Quality Control Department Functions

Quality control departments play a pivotal role in producing good quality textile products in a textile organizational/mill/factory. They perform functions such as shown in Figure 10.2 [9]. Assigning control responsibilities is concerned with defining and assigning control responsibilities for items such as checking, control measurements, and weighing of waste throughout the factory/mill. Training mill/factory manpower is concerned with planning and providing appropriate quality-related training to factory/mill personnel. Establishing and maintaining the testing laboratory is concerned with setting up and maintaining the testing laboratory with appropriate equipment and qualified manpower. Ensuring prompt execution of corrective measures is concerned with coordinating corrective actions in such a manner that take minimum time between the discovery of faulty operation and corrective measure.

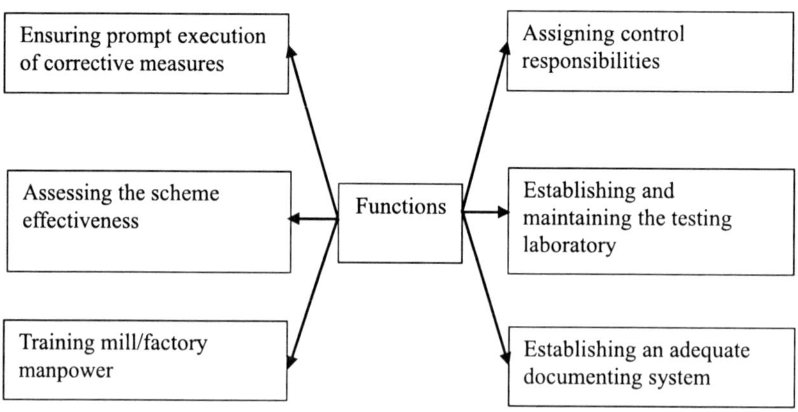

Figure 10.2. Main functions of a textile quality control department

Assessing the scheme effectiveness is concerned with regularly reviewing the scheme and making changes as considered appropriate. Establishing an adequate documenting system is concerned with designing forms for purposes such as recording measurements, calculations, summaries of measurement changes with time, and control charts.

10.5 Textile Test Methods

Over the years a large number of test methods have been developed to determine various different aspects of textiles that may directly or indirectly affect textile quality [10]. Some of these aspects are presented in Table 10.1 [10]. Five test methods considered, directly or indirectly, useful in textile quality control work are briefly described below.

Table 10.1. Some aspects of textiles determined by the textile test methods

No.	Test method for determining (Textile aspect)
1	Moisture in textiles
2	Length of yarns and threads
3	Breaking strength of fabrics
4	Breaking strength of yarns
5	Tearing strength
6	Flame resistance
7	Water resistance
8	Air permeability
9	Resistance to pilling
10	Yarn crimp

10.5.1 Test Method I

This method is concerned with determining the conditioned mass of fabrics by taking specimens of known dimensions from fabric in moisture equilibrium with the standard atmosphere. The method calls for using a balance capable of determining specimen mass with an accuracy of $\pm 0.1\%$ as well as a rigid scale graduated in millimetres and centimetres. Additional information on the method is available in Ref. [10].

10.5.2 Test Method II

This method is concerned with determining the tearing strength of woven fabrics by measuring the maximum force observed in the propagation of a tear across the fabric when the force applied is parallel to the yarns ruptured in the tear. The method calls for using a suitable recording tensile testing machine of the inertia less or pendulum type. Additional information on the method is available in Ref. [10].

10.5.3 Test Method III

This method is concerned with determining the rate at which a strip of fabric burns when a flame is applied to a vertical specimen's lower edge until it ignites. The time is estimated for the upper portion of the flame to travel a distance of 650 mm up the fabric strip. The method uses a small burner, a stop watch, and a shield about 300-mm wide, 300-mm deep, and 1.2-m high, with an open top and a sliding/hinged glass front for protecting the specimen from draft. This method is described in detail in Ref. [10].

10.5.4 Test Method IV

This method is concerned with determining the amount of crimp in yarns taken from woven fabric. The difference between a measured length of fabric and the straightened length of a yarn subtracted from it is estimated and presented as a percentage of the fabric's measured length. The apparatus required for this test are a rigid scale and a tensioning device (*e. g.*, a twist tester). This method is described in detail in Ref. [10].

10.5.5 Test Method V

This method is concerned with determining the tearing strength of woven fabrics by the single-rip approach. In this method the force needed for propagating a single-rip tear through a fabric, is estimated and then its maximum values in successive equal size tearing intervals are averaged. The method requires a suitable recording tensile testing machine. Additional information on this method is available in Ref. [10].

10.6 Quality Control in Spinning and Fabric Manufacture

The main objective in spinning is to manufacture a yarn of defined count and quality at the minimum cost [9]. The quality should be adequate for ensuring that the yarn performs effectively in subsequent processes as well as the end product is with in acceptable limits. Factors such as the yarn's uniformity, tensile strength, elongation, and freedom from imperfections determine quality. The relative importance of factors such as these depends on subsequent processes and the end product. The basic characteristics of spinning are count and count uniformity since yarn is designated by count and both these characteristics affect strength and strength variability, the performance in subsequent processes, and the fabric appearance.

Waste is an important factor in spinning economics and is becoming increasingly important due to factors such as the need for large capital investment in machinery, global competition, and the growing costs of raw material and labour. The amount of waste may increase with either decreases/increases in quality, or it may be totally independent of quality. Some of the important factors for excess waste could be inadequate operating procedures, poor management/supervision, poor training, operator carelessness, and lack of operator skill.

The main goals of quality control in fabric/cloth manufacturing are to achieve the specified level of quality with minimum waste, and maintain optimum level of machine and labour productivity so that the profit is at maximum [9]. Fabric defects in weaving result from defects in the preparatory process, poor work practices, wrong loom settings, and end breakages (because of poor loom maintenance).

The fabric mechanical properties are appeared to be mainly influenced by the stitch length and the yarn shear and bending. As the long-term retention properties such as abrasion resistance and lack of pilling involve so many factors, an examination of the finished fabric is the only reasonable test. Nonetheless, the three factors that should be tested are yarn variables, process variables, and fabric variables.

Yarn variables involve checks on count, checks on knots, slobs, and thin spots, and measurement of yarn irregularity. Furthermore, to prevent press-offs and broken loops, yarn strength and strength variability need to be controlled. Process variables basically involve daily checks on input tension and stitch length and visual inspection of fabric for correct pattern selector operation.

Fabric variables involve tests for abrasion, pilling, and dimensional stability and inspection for irregular and dropped stitches, and rough dimensional checks. All in all, some organizations or factories rely totally on these checks for quality control.

10.7 Quality Control in Finishing and in the Clothing Industry

Quality control at the finishing stage involves many different functions. They may be grouped under four distinct categories as shown in Figure 10.3 [9].

Two of these categories are discussed below. Control of raw materials is essential, since they are often purchased from a specific source on the basis of price rather than on quality. It is very important to carry out a careful testing of a raw material where its behaviour in processing is crucial and its cost is a major element in the cost of the final product.

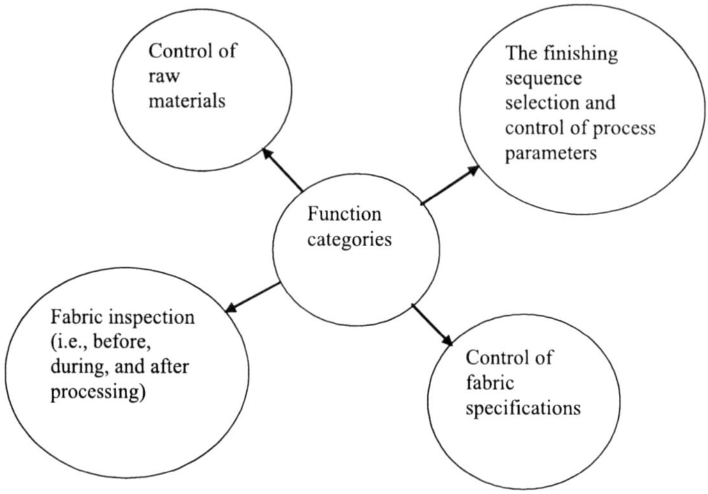

Figure 10.3. Quality control function categories at the finishing stage

The selection of the appropriate processing sequence and processing parameters depends on many parameters, including the type of fabric, the properties required, and the fibres used. To suit a specific fibre blend, processes may have to be modified or changed and more effective control may be required to minimize faults. Additional features of quality control in finishing are good machine maintenance, cleanliness, and tidiness.

Quality control in the clothing industry is not very clear-cut due to various reasons including yarn properties, a wide variety of fabrics to be handled by clothing manufacturers, the dependence of quality on the fibres used, the manufacturing parameters, and the finishing variables [9]. Nonetheless, quality control may be divided into three distinct areas: performance testing, acceptance testing, and product inspection.

Performance testing involves particular tests on properties critical to specific types of fabrics. Some examples of these tests are fabric-to-fabric adhesion in fusible interlinings, inflammability of children's garments, air permeability in wind-proof fabrics, and shower-proofing for rainwear.

Acceptance testing incorporates the testing of all types of raw materials used, including elasticized waist-band fabric, tapes, pocketing, linings, padding, sewing threads, interlinings, stiffening, and basic fabric and auxiliaries such as zippers, press studs, buttons, hooks, and eyes.

Product inspection is concerned with removing processing faults, and it ensures that no further work is done on garments or items already identified as faulty and the final appropriate inspection prevents their sale. The degree of inspection to be performed depends on factors such as the type of garment, the specified quality, and the price range. All in all, the application of the quality control concept in the clothing industry is very challenging basically due to two reasons: the variability of input raw materials and the large range and short production runs of the product [9].

10.8 Organizations that Issue Textile Standards

There are many organizations around the world that issue textile-related standards, directly or indirectly, useful to quality control in the textile industry. Some of these organizations along with their addresses are as follow [9]:

- International Standards Organization (ISO)

  1 Rue de Varembe
  1211 Geneva 20
  Switzerland

- International Wool Textile Organization (IWTO)

  Hastlegate, Bradford
  Yorkshire BD1 1DE
  Untied Kingdom

- National Bureau of Standards

  Washington, D.C.
  USA

- British Standards Institution (BSI)

  Textile Division
  10 Blackfriars Street
  Manchester M3 5DR
  United Kingdom

- Council of the European Economic Community

  200 rue de la Loi
  B-1040 Brussels
  Belgium

- American Society for Testing and Materials (ASTM)

  1916 Race Street
  Philadelphia, PA 19103
  USA

- Pan American Standards Commission

  c/o Argentine Standards Institute (IRAM)
  Chile 1192
  Buenos Aires
  Argentine

- Canvas Products Association International

  600 Endicott Buildings
  St. Paul, MN 55101
  USA

10.9 Problems

1. Write an essay on quality control in the textile industry.
2. List at least seven quality-related issues associated with textiles.
3. Discuss major quality problems experienced in apparel.
4. What is a yarn?
5. What are the main functions of a textile quality control department?
6. Discuss quality control in the area of spinning.
7. Discuss quality control in the following two areas:
   - Finishing
   - Fabric manufacture
8. Write an essay on fibers.
9. List at least eight different aspects of textiles determined by the textile test methods.
10. What are the important organizations useful for obtaining textile quality-related standards?

References

1 Economic Developments in the Textile Industry, A Report by the American Textile Manufacturers Institute and the Office of the Chief Economist, Washington, D.C., September, 2000.
2 Clapp, T.G., Godfrey, A.B., Greeson, D., Jonson, R.H., Rich, C., Seastrunk, C., Quality Initiatives Reshape the Textile Industry, Quality Digest, October 2001. Available Online at http://www.qualitydigest.com/oct-01/html/textile.html.
3 Siekman, P., The Big Myth About U.S. Manufacturing, Fortune Magazine, October 2, 2000.
4 Winners Showcase, National Institute of Standards and Technology, Washington, D.C., August 25, 2000. Available online at http://www.quality.nist.gov/winners/milliken.html.
5 Winchester, S.C., Total Quality Management in Textiles, Journal of the Textile Institute, Vol. 85, No. 4, 1994, pp. 445–459.
6 Page, H.S., A Quality Strategy for the 80's, Quality Progress, Vol. 16, No. 11, 1983, pp. 16–21.
7 Lees, G., Quality Philosophies and Practices in the Garment Business, Quality Assurance, Vol. 11, No. 1, 1985, pp. 21–24.
8 Manual on Instrumentation and Quality Control in the Textile Industry, Development and Transfer of Technology Series No. 4, United Nations, New York, 1978.
9 Quality Control in the Textile Industry, United Nations Publication Sales No. E.72.II.B.24, United Nations, New York, 1972.
10 National Standard of Canada No. CAN2-4.2-M77, Textile Test Methods, Published by the Canadian Government Specifications Board, Ottawa, Ontario, Canada, 1977.

11

Quality Control in the Food Industry

11.1 Introduction

The food industry is a huge global business, for example, people in the United States alone eat about 870 million meals a day [1]. In this industry, quality is usually an integrated measure of purity, texture, appearance, flavour, workmanship, and color. Quality is becoming an important issue due to various factors including food borne disease, caused by improper food handling or storage. For example, during the period 1993 to 1997, in the United States a total of 2,751 outbreaks of food borne disease involving around 86,000 people, were reported [2]. The main causes for the disease were identified as bacteria, viruses, parasites, and chemical agents.

The history of laws, directly or indirectly, concerned with food quality can be traced back to 1202 when King John of England proclaimed the first English food law, the Assize of Bread, which prohibited adulteration of bread with such ingredients as beans or peas. Nonetheless, some of the main objectives of quality control in the food industry are as follows:

- To assure that food laws are complied with in an effective manner.
- To protect people from dangers (*e. g.*, contaminated foods) and ensure that they get the proper quality and weight as per payments.
- To provide protection to the business from cheating by its suppliers, damage to equipment (*e. g.*, stones in raw materials), and false accusations by customers, suppliers, or middlemen.

This chapter presents various important aspects of quality control in the food industry.

11.2 Factors Affecting Food Quality and Basic Elements of a Food Quality Assurance Program

There are many factors responsible for poor quality food. The major factors responsible for significant quality changes are listed below [3]:

- Wrong temperatures and timing
- Poor packaging
- Inadequate machine maintenance program
- Wrong pre-cooking, cooking, and post-cooking approaches or methods
- Poor ware washing
- Poor sanitation
- Presence of pesticides
- Incompatible water conditions
- Presence of vermin
- Incorrect formulations, stemming from wrong weight of the food, or its elements/components
- Spoilage due to chemical, biochemical, microbiological, or physical factors

All in all, any of the above factors can contribute to poor food quality as well as effect changes that could be evident in the food's appearance, texture, consistency, and flavour.

Ten basis elements of a food quality assurance program are shown in Figure 11.1 [3]. Some of these elements are described below.

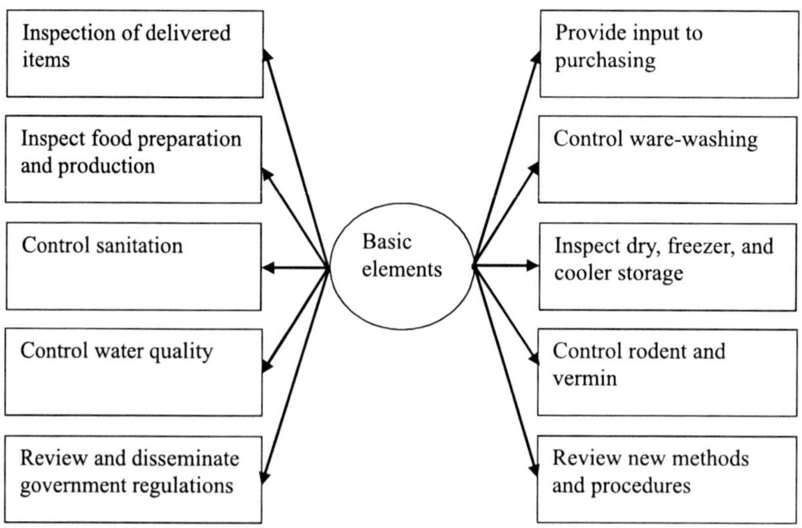

Figure 11.1. Basic elements of a food quality assurance program

The inspection of delivered items is concerned with actions such as follows:

- Conducting comparison tests to compare the delivered items with purchase specifications.
- Inspecting the product for visual signs of contamination and freshness.
- Recording the temperature of the frozen food.
- Determining product weight.
- Recording pack data and product code.
- Checking label nomenclature for conformity to labelling standards.
- Checking canned merchandise for dents and "swells".

The element "provide input to purchasing" is concerned with items such as developing procedures for test panels and cooking tests, and establishing specifications and formulations for each food and beverage item to be purchased.

The element "inspect food preparation and production" is concerned with actions such as follows:

- Testing quality of finished food, beverages, garnishing, and plating.
- Checking efficiency of all cooking equipment with respect to temperature, timing, and physical condition.
- Controlling sanitation to eliminate problems of off-tastes, off-flavours, and food spoilage.
- Ensuring against over-production.
- Updating and reviewing recipe cards and other data useful for formulating and preparing food.

The element "inspect dry, freezer, and cooler storage" is concerned with actions such as evaluating storage temperature, establishing procedures for proper storage of left-over food, controlling sanitation, and developing orderly stacking procedures. The element "control ware-washing" is concerned with assuring the total removal of soap, grease, and soil residues.

The element "control sanitation" is concerned with sanitation control for refuse collection and disposal area. The element "review new methods and procedures" is concerned with reviewing new procedures and approaches pertaining to food production, packaging, and handling. The element "review and disseminate government regulations" is concerned with reviewing and disseminating local, state, and federal government health-related regulations pertinent to the establishment.

11.3 Total Quality Management Tools for Application in the Food Industry

As total quality management (TQM) is based on many ideas, it means thinking about quality in terms of all functions of the food processing or other organization [3]. TQM makes use of many tools to successfully achieve the desired objectives.

The TQM tools considered useful for application in the food industry can be divided into two groups (*i. e.*, Group I and Group II). Group I tools are concerned with analyzing and interpreting numerical data whereas Group II with management and planning.

Some of the most useful tools belonging to the Group I Category are briefly discussed below [4–7].

- **Histogram.** Histogram can be used to summarize and display the distribution of a given food process dataset. A histogram is quite useful to answers questions such as what is the most frequent system response? And what distribution (*i. e.*, shape, center, and variation) do the data have? Additional information on histograms is available in Refs. [4–6].
- **Flowchart.** Flowcharts are an excellent project development and documentation tool. A flowchart visually records the decisions, steps, and actions of a given service or manufacturing operation as well as defines the system and its associated pivotal points, activities, and role performances. Additional information on flowcharts is available in Refs [4–6].
- **Control chart.** Control charts are one of the most technically sophisticated methods of statistical quality control, and they can be used to identify statistically significant changes that may happen in a food-related process. A control chart may simply be described as a graphic presentation of data collected over a period of time, which shows upper and lower control limits for the process to be controlled. Thus, each control chart is comprised of three lines: upper control limit (UCL), lower control limit (LCL), and the center line. In the food industry, control charts are often used for net weight control. Control charts are described in detail in Refs. [1, 5, 8].
- **Pareto diagram.** A pareto diagram a kind of frequency chart in which bars are arranged in descending order (*i. e.*, from left to right) and which provides order to activity. A Pareto diagram is used to highlight areas for a concerted effort (*i. e.*, to decide what steps need to be taken for quality improvement). More specifically, a Pareto diagram is a valuable tool for seeking answers to questions such as what 20% of sources are responsible for 80% of the problems? And what are the most pressing issues facing a business or team? Additional information on this method is available in Refs. [4–7].
- **Scatter diagram.** A scatter diagram is quite similar to a line graph, but with one exception, *i. e.*, the data values are plotted without a connecting line drawn between them. The scatter diagram is used to study all possible relationships between two given variables. Although it cannot prove that one variable causes the other, but it does indicate the existence of a relationship as well as that relationship's strength. Additional information on this approach is available in Refs. [4–6].

Similarly, five of the most useful tools belonging to the Group II category are briefly described below [4–6].

- **Inter-relationship diagraph.** An inter-relationship diagram is used to find solutions to problems having complex causal relationships. An inter-relationship diagraph may simply be described as a process that allows for multidirectional rather than linear thinking to be used. All in all, the inter-relationship diagraph is an excellent tool for untangling and finding the logical relations among the intertwined causes and effects and is described in detail in Refs. [4–6, 9].
- **Affinity diagram.** An affinity diagram is a process used by a team or group for collecting and organizing opinions, issues, ideas, etc. from a raw list into classifications of similar thoughts that make sense and can be handled more easily or effectively. Some of the situations when the affinity diagram can be used are when pre-existing ideas need to be clarified or overcome, when there is a definite need to create unity within a group, and when thoughts/facts are uncertain and need to be organized. Additional information on the affinity diagram is available in Refs. [4–6, 10].
- **Tree diagram.** Tree diagrams are used for mapping out a full range of tasks and paths that must be performed in order to accomplish a specified primary goal and associated subgoals. The tree diagram permits breaking any broad objectives or goals, graphically, into increasing levels of detailed actions that must or could be accomplished to successfully achieve the specified goals. This method is described in detail in Ref. [10].
- **Process decision program chart.** Process decicsion program chart is a powerful approach that graphically displays various alternatives and contingencies to a given problem, which can be determined in advance to chose a strategy for handling them. The process decision program chart can be used for purposes such as implementing countermeasures to minimize non-conformities in the manufacturing process, exploring all possible contingencies that could occur in the implementation of any new or untried risky plan, and establishing an implementation plan for management by objectives [4]. Additional information on this method is available in Refs. [4–6].
- **Matrix diagram.** Matrix diagrams are used to visually examine the relationship between data sets. The diagram is composed of a number of rows and columns whose intersections are compared to determine the nature and strength of the problem under consideration. This permits the user to come up with the most promising ideas, analyze the relationship or its absence at the intersection, and determine a useful way of pursuing the problem-solving approach. The matrix diagram is described in detail in Refs. [4–6].

11.4 Hazard Analysis and Critical Control Points (HACCP) Concept

Today, the Hazard Analysis and Critical Control Points (HACCP) concept is widely used in the food industry. It was developed in the 1960s by the Pillsbury Corporation in conjunction with the National Aeronautics and Space Administration

(NASA) and the US Army Natick Laboratories to ensure the safety of food for astronauts [4, 11–13]. Nowadays, HACCP has clearly become a technical management program in which food safety is addressed by controlling chemical, physical, and biological hazards in all areas of the food industry (*i. e.*, from growing, harvesting, processing, and distribution to preparing all types of food for consumption).

In developing an HACCP program, there are five basic tasks that must be accomplished successfully prior to the application of the HACCP principles to a certain product and process [4, 14]. These five tasks are as follows [4, 14]:

- Forming the HACCP team with appropriate individuals with appropriate expertise in required areas.
- Describing the food product and its distribution.
- Describing the food product's intended use and its potential consumers.
- Developing a flow diagram that describes the food product's manufacturing process.
- Verifying the flow diagram.

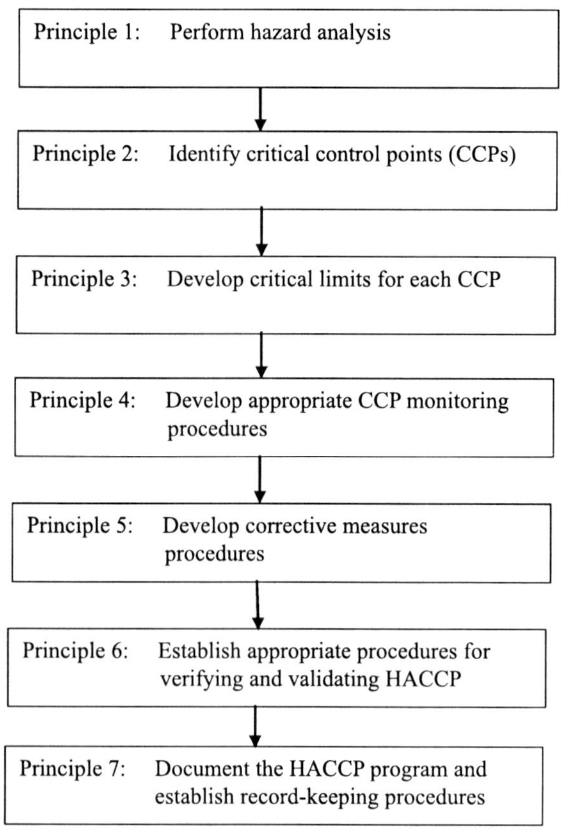

Figure 11.2. The seven HACCP principles

After the completion of the above five tasks, the seven principles of HACCP shown in Figure 11.2 are applied. Each of these principles is described in detail in Ref. [4].

11.4.1 HACCP Benefits

There are many benefits of the HACCP concept. Some of the important ones are as follows [4, 15]:

- HACCP is a useful tool to place responsibility for ensuring food safety on food manufacturers or distributors.
- HACCP is a useful tool that focuses on identifying and preventing hazards from contaminating food.
- HACCP is a useful tool that permits more effective and efficient government monitoring, primarily because of its record keeping, allows investigators to see how well an organization is complying with food safety-related laws over a period of time.
- HACCP is a useful tool for reducing barriers to international trade.
- HACCP is a useful tool that helps food companies to compete more effectively in the global market.

11.5 Fruits and Vegetables Quality

As the health benefits associated with regular consumption of fresh fruits and vegetables have been clearly demonstrated and encouraged by nutrition and health authorities, the increase in consumption of these products has been an important factor of the greater emphasis on their quality. There are many factors that affect the quality of fruits and vegetables. Most of these factors are shown in Figure 11.3 [16]. Each of these factors is described in detail in Ref. [16].

11.5.1 Main Causes of Post-harvest Losses and Poor Quality for Various Categories of Fruits and Vegetables

This section presents important causes of post-harvest losses and poor quality for the following five categories of fruits and vegetables [17]:

- **Category I: Root Vegetables.** This category includes vegetables such as beets, onions, sweet potatos, garlics, potatos, and carrots. Their main causes of post-harvest losses and poor quality are water loss, mechanical injuries, chilling injury, improper curing, decay, and sprouting.
- **Category II: Flower Vegetables.** This category includes vegetables such as cauliflower, broccoli, and artichokes. Their main causes of post-harvest losses

and poor quality are discoloration, water loss, mechanical injuries, and abscission of florets.

- **Category III: Leafy Vegetables.** This category includes vegetables such as spinach, cabbage, lettuce, green onions, and chard. Their main causes of post-harvest losses and poor quality are mechanical injuries, water loss, decay, loss of green color, and relatively high respiration rates.
- **Category IV: Immature Fruit Vegetables.** This category includes vegetables such as cucumbers, okra, peppers, eggplant, snap beans, and squash. Their main causes of post-harvest losses and poor quality are water loss, chilling injury, decay, bruising and other mechanical injuries, and over-maturity at harvest.
- **Category V: Mature Fruit Produce.** This category includes fruits such as apples, melons, bananas, grapes, tomatoes, stone fruit, and mangoes. Their main causes of post-harvest losses and poor quality are water loss, bruising, chilling injury, decay, compositional changes, and over-ripeness at harvest.

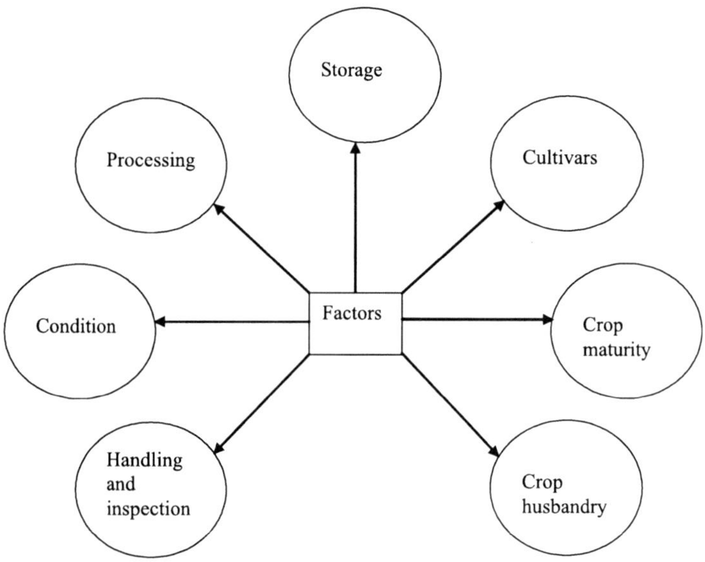

Figure 11.3. Factors affecting quality of fruits and vegetables

11.6 Vending Machine Food Quality

A vending machine may simply be described as a self-service device that upon insertion of coins, tokens, or debit cards automatically dispenses unit servings of food either in packaged form or in bulk. Today, vending machines are widely used throughout the world. For example, in Japan alone, there are around five million vending machines covering a wide range of products and services. Three major areas for quality control of food and beverage vending machines are shown in

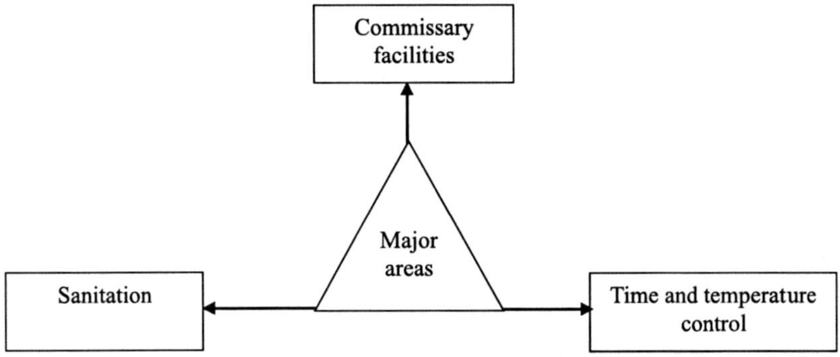

Figure 11.4. Major areas for quality control of food and beverage vending equipment

Figure 11.4 [18]. These are sanitation, time and temperature control, and commissary facilities.

Sanitation is basically concerned with cleaning and sanitizing all components of a vending machine that come into contact with food, in such a manner that prevents the contamination of food being served from the machine. Food-contact parts of a vending machine cannot be effectively cleaned with simply a rag and a dash of water, because of the machine complexity. This cleaning requires a careful attention and inspection. In addition, a professional job cannot be accomplished without proper sanitation equipment.

Although the need for sanitation items depends on type of machine, company procedures, and machine location, the suggested items for the sanitation kit are three buckets (*i. e.*, for the detergent solution, sanitizing solution, and hot water rinse), hand mops and sponges, insecticide spray, hand scrapers and soft wire brushes, cleaning cloths and paper and cloth towelling with high wet strength properties, brushes of various sizes, flashlight, spare water strainer and filter cartridge, spare tubing for replacement purposes, spare polyethylene waste bags, and detergents, approved sanitizers, urn cleaner and cleaner spray in bomb or bottle [18].

Time and temperature control calls for establishing a rigid program of time and temperature control of perishable foods and beverages. The safe recommended temperatures by the Public Health Department of the National Automatic Merchandising Association (USA), are 45°F (8°C) or lower for cold food and 140°F (60°C) or higher for hot food [18].

Commissary facilities are concerned with preparing foods such as salads, casseroles, stews, and sandwiches as well as performing the role of a storage and distribution center. The commissary operation and quality control procedures are basically identical to those needed for any other food service facility. However, their differences are quite apparent and include containerization of salads and other prepared foods, wrapping of pastries and sandwiches, and transporting all these items to vending machines.

11.6.1 Important Points in the Quality Control of Vended Hot Chocolate

Some important points in the quality control of vended hot chocolate are as follows [18]:

- Maintenance of water temperature at 200°F ± 5° (94°C ±3°).
- Avoiding the overload of hopper.
- Checking weekly the product quantity or throw.
- Flushing the mixing chamber after servicing a machine.
- Checking the hopper cover for ensuring snug and moisture-free fit to prevent the growth of surface mold and bacteria

11.6.2 Quality Control Factors for Soft Drink Vending Machines

Some of the quality control factors for soft drink vending machines are shown in Figure 11.5 [18].

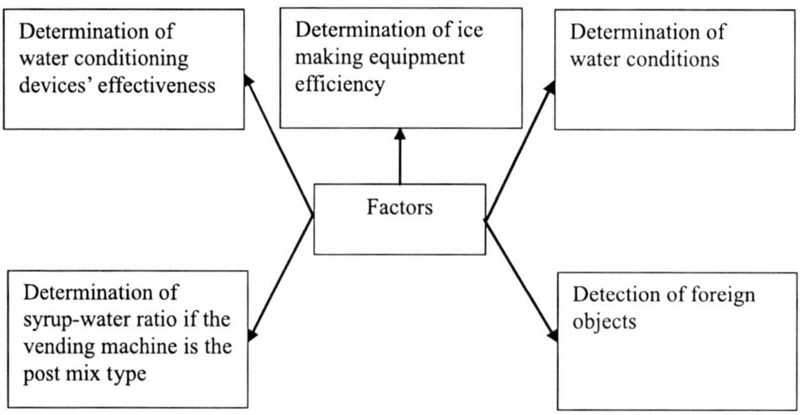

Figure 11.5. Quality control factors for soft drink vending machines

11.7 Food Processing Industry Quality Guidelines

Food processing industry quality guidelines are based upon ANSI Z1.15, an American National Standards Institute (ANSI) standard for establishing quality control systems in hardware manufacturing [1, 19]. More specifically, in the 1980s a committee of food quality experts modified this standard for use by the food processing industry. The modified version covers seven areas shown in Figure 11.5 [1]. These are administration, design assurance and design change

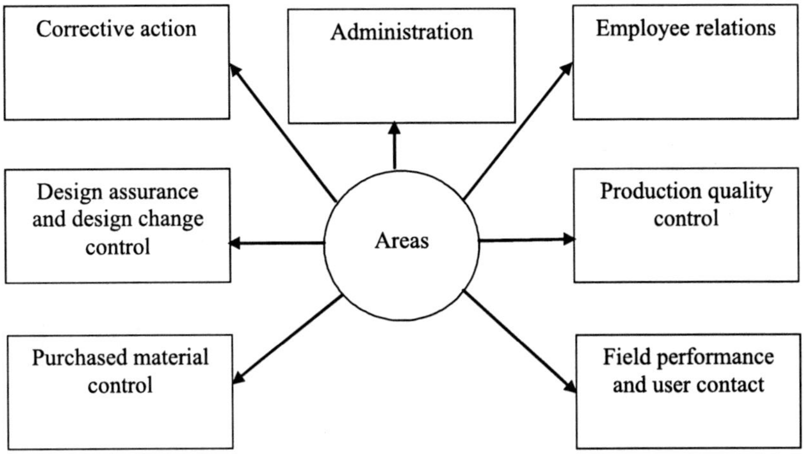

Figure 11.6. Areas covered by the quality standard for use by the food processing industry

control, purchased material control, production quality control, field performance and user contact, employee relations, and corrective action.

Administration includes items such as quality system, objectives, quality policy, planning quality manual, responsibility, reporting, quality cost management, and quality system audits. Each of these items is covered in significant depth. For example, the quality system covers items such as ingredients, sanitation, distribution, packaging, storage practices, pest management/control, vendor/contract processors relations, user contacts, complaint handling and analysis, shelf life, processing, and finished product.

Design assurance and design change control contains a total of twelve subsections concerned with design review, concept definition, market readiness reviews, etc. The purchased material control provides a summary of supplier certification requirements such as system requirements, specifications, assistance to suppliers, and facility inspection.

Production quality control contains a total of twenty four detailed requirements under many subheadings: finished product inspection, planning and controlling the process, quality information, product and container marking, and handling, storage, and shipping.

Field performance and user contact includes items such as complaints and analysis, product objective, advertising, and acceptance surveys. The employee relations area includes selection, motivation, and training. Finally, corrective action covers items such as diction, documentation, and incorporating change.

11.8 Problems

1. Write an essay on quality control in the food industry.
2. List at least of ten the most important factors responsible for significant quality changes.
3. What are the basic elements of a food quality assurance program? Discuss at least five of them.
4. Discuss at least five total quality management tools useful for application in the food industry.
5. What are the seven principles associated with the hazard analysis and critical control points (HACCP) concept?
6. What are the advantages of HACCP?
7. What are the factors that affect the quality of fruits and vegetables?
8. Discuss three major areas for quality control of food and beverage vending equipment.
9. List important points in the quality control of vended hot chocolate.
10. Discuss main causes of post-harvest losses and poor quality for at least three different categories of fruits and vegetables.

References

1 Hubbard, M.R., *Statistical Quality Control for the Food Industry*, Kluwer Academic/Plenum Publishers, New York, 2003.
2 Ong, K.G., Puckett, L.G., Sharma, B.V., Loiselle, M., Grimes, C.A., Bachas, L.G., *Wireless, Passive, Resonant-Circuit Sensors for Monitoring Food Quality*, Proceedings of SPIE, Vol. 4575, 2002, pp. 150–159.
3 Thorner, M.E., Manning, P.B., *Quality Control in Food Service*, The AVI Publishing Company, Westport, Connecticut, 1976.
4 Vasconcellos, J.A., *Quality Assurance for the Food Industry*, CRC Press, Boca Raton, Florida, 2004.
5 Mears, P., *Quality Improvement Tools and Techniques*, McGraw Hill Book Company, New York, 1995.
6 Kanji, G.K., Asher, M., *100 Methods for Total Quality Management*, Sage Publications Ltd., London, 1996.
7 Dhillon, B.S., *Advanced Design Concepts for Engineers*, Technomic Publishing Company, Lancaster, Pennsylvania, 1998.
8 Dhillon, B.S., *Reliability, Quality, and Safety for Engineers*, CRC Press, Boca Raton, Florida, 2005.
9 Pyzdek, T., *The Six Sigma Handbook*, McGraw Hill Book Company, New York, 2003.
10 Mizuno, S., Editor, *Management of Quality Improvement: The Seven New QC Tools*, Productivity Press, Cambridge, Massachusetts, 1988.

11 Bauman, H.E., *The Hazard Analyses Critical Control Concept, in Food Protection Technology*, edited by C.W., Felix, Lewis Publishers, Chelsea, Michigan, 1987, pp. 175–179.

12 Simonson, B., Bryan, F.L., Christian, J.H.B., Roberts, T.A., Tompkin, R.B., Silliker, J.H., *Prevention and Control of Food borne Salmonellosis Through Application of Hazard Analysis Critical Control Point (HACCP)*, International Journal of Food Microbiology, Vol. 4, 1987, pp. 227–247.

13 Shaw, S., Rose, S.A., *New Food Legislation and the Role of Quality Assurance*, Quality Forum, Vol. 17, No. 4, 1991, pp. 151–155.

14 The Quality Auditor's *HACCP Handbook*, ASQ Quality Press, Milwaukee, Wisconsin, 2002.

15 Pierson, M.D., Corlett, D.A., *HACCP Principles and Applications*, Van Nostrand Reinhold Company, New York, 1992.

16 Arthey, D., *Quality Control of Fruits and Vegetables and Their Products*, in Quality Control in the Food Industry edited by S.M. Herschdoerfer, Academic Press, London, 1986, pp. 217–260.

17 *Improving the Safety and Quality of Fresh Fruits and Vegetables: A Training Manual for Trainers*, published by the Joint Institute for Food Safety and Applied Nutrition (JIFSAN), University of Maryland, College Park, Maryland, USA, 2002.

18 Thorner, M.E., Manning, P.B., *Quality Control in Food Service*, The AVI Publishing Company, Westport, Connecticut, 1976.

19 ANSI Z1.15, *Generic Guidelines for Quality Systems*, American National Standards Institute, New York, 1980.

Appendix

Bibliography: Literature on Applied Reliability and Quality

A.1 Introduction

Over the years, a large number of publications, directly or indirectly, related to various areas of applied reliability and quality have appeared in the form of journal articles, conference proceedings articles, books, etc. This appendix presents an extensive list of such publications [1–857] on all the applied reliability and quality areas covered in the book. These publications are separated into each of these applied areas: quality in healthcare ([1–50] for the period 1989–2005), Internet reliability ([51–117] for the period 1995–2004), quality control in the food industry ([118–203] for the period 1979–2006), quality control in the textile industry ([204–255] for the period 1960–2005), software quality ([256–560] for the period 1990–2005), robot reliability ([561–636] for the period 1993–2004), power system reliability ([637–836] for the period 1990–2006), and medical equipment reliability ([837–857] for the period 2000–2005).

The main objective of this listing is to provide readers with sources for obtaining additional information on applied reliability and quality.

A.2 Publications

A.2.1 Quality in Healthcare

1 Al-Assaf, A.F., Schmele, J.A., The Textbook of Total Quality in Healthcare, St. Lucie Press, Delray Beach, Florida, 1993.

2 Anon, "The Quality of the NHS," *Quality World,* Vol. 29, No. 2, 2003, p. 14.

3 Batalden, P., "Deming offers much to Healthcare Workers," *Quality Progress,* Vol. 35, No. 12, 2002, pp. 10–12.

4 Bell, R., Krivich, M.J., How to Use Patient Satisfaction Data to Improve Healthcare Quality, ASQ Quality Press, Milwaukee, Wisconsin, U. S. A., 2000.

5 Ben-Zvi, S., "Quality Assurance in Transition," *Biomedical Instrumentation & Technology,* Vol. 23, No. 1, 1989, pp. 27–33.

6 Berndt, D.J., Fisher, J.W., Hevner, A.R., "Healthcare Data Warehousing and Quality Assurance," *Computer,* Vol. 34, No. 12, 2001, pp. 56–65.

7 Beuscart, R.J., Alao, O.O., Brunetaud, J.M., "Health Telematics: A Challenge for Healthcare Quality," *Proceedings of the 23rd Annual International Conference of the IEEE Engineering in Medicine and Biology Society,* 2001, Vol. 4, pp. 4105–4107.

8 Bombino, A., "Lessons Learned in Applying QFD at Baxter Healthcare," *Proceedings of the NEPCON West '92 Conference,* 1992, Vol. 2, pp. 877–880.

9 Bowling, P., Berglund, R., "HIPAA: Where Healthcare and Software Quality Meet," *Proceedings of the 57th Annual Quality Congress: Expanding Horizons,* 2003, pp. 271–281.

10 Brown, P.J.B., and Warmington, V., "Data Quality Probes – Exploiting and Improving the Quality of Electronic Patient Record Data and Patient Care," *International Journal of Medical Informatics,* Vol. 68, No. 1–3, 2002, pp. 91–98.

11 Caroselli, M., Edison, L., Quality Care: Prescriptions for Injecting Quality into Healthcare Systems, St. Lucie Press, Boca Raton, Florida, 1997.

12 Chae, Y.M., Kim, H.S., Tark, K.C., "Analysis of Healthcare Quality Indicator Using Data Mining and Decision Support System," *Expert Systems with Applications,* Vol. 24, No. 2, 2003, pp. 167–172.

13 Chaplin, E., "Customer Driven Healthcare Comprehensive Quality Function Deployment," *Proceedings of the 56th Annual Quality Congress,* 2002, pp. 767–781.

14 Elkin, P.L., Brown, S.H., Carter, J., "Guideline and Quality Indicators for Development, Purchase and Use of Controlled Health Vocabularies," *International Journal of Medical Informatics,* Vol. 68, No. 1–3, 2002, pp. 175–186.

15 Fried, R.A., "TQM in the Medical School: A Report from the Field," *Proceedings of the 45th Annual Quality Congress Transactions,* 1991, pp. 113–118.

16 Frings, G.W., Grant, L., "Who Moved my Sigma ...Effective Implementation of the Six Sigma Methodology to Hospitals," *Quality and Reliability Engineering International,* Vol. 21, No. 3, 2005, pp. 311–328.

17 Ghahramani, B., "An Internet based Total Quality Management System," *Proceedings of the 34th Annual Meeting of the Decision Sciences Institute,* 2003, pp. 345–349.

18 Hogl, O., Muller, M., Stoyan, H., "On Supporting Medical Quality with Intelligent Data Mining," *Proceedings of the 34th IEEE Annual Hawaii International Conference on System Sciences,* 2001, pp. 141–145.

19 Iz, P.H., Warren, J., Sokol, L., "Data Mining for Healthcare Quality, Efficiency, and Practice Support," *Proceedings of the IEEE 34th Annual Hawaii International Conference on System Sciences,* 2001, pp. 147–152.

20 Khorramshahgol, R., Al-Barmil, J., Stallings, R.W., "TQM in Hospitals and the Role of Information Systems in Providing Quality Services," *Proceedings of the IEEE Annual International Engineering Management Conference,* 1995, pp. 196–199.

21 Kimberly, J.R., Minvielle, E., The Quality Imperative: Measurement and Management of Quality in Healthcare, Imperial College Press, London, 2000.

22 King, R., "Six Sigma and its Application in Healthcare," *Proceedings of the ASQ's 57th Annual Quality Congress: Expanding Horizons,* 2003, pp. 39–47.

23 Kirk, R., Healthcare Quality & Productivity: Practical Management Tools, Aspen Publishers, Rockville, Maryland, 1988.
24 Labovitz, G.H., "Total-quality Health Care Revolution," *Quality Progress,* Vol. 24, No. 9, 1991, pp. 45–47.
25 Larimer, M.L., Bergquist, T.M., "A Comparison of Three Quality Improvement Systems for Health Care," *Proceedings of the 34th Annual Meeting of the Decision Sciences Institute,* 2003, pp. 1843–1848.
26 Le Duff, F., Daniel, S., Kamendje, B., "Monitoring Incident Report in the Healthcare Process to Improve Quality in Hospitals," *International Journal of Medical Informatics,* Vol. 74, No. 2–4, 2005, pp. 111–117.
27 Marshall, D., "Health Care Quality: It is not 'Job One'!", *Proceedings of the 56th ASQ Annual Quality Congress*, 2002, pp. 83–91.
28 Marszalek-Gaucher, E., Total Quality in Healthcare: From Theory to Practice, Jossey-Bass Publishers, San Francisco, 1993.
29 McDaniel, J.G., "Improving System Quality Through Software Evaluation," *Computers in Biology and Medicine,* Vol. 32, No. 3, 2002, pp. 127–140.
30 Merry, M.D., "Healthcare's Need for Revolutionary Change," *Quality Progress,* Vol. 36, No. 9, 2003, pp. 31–35.
31 Moraes, L., Garcia, R., "Contribution for the Functionality and the Safety in Magnetic Resonance: An Approach for the Imaging Quality," *Proceedings of the 25th Annual International Conference of the IEEE Engineering in Medicine and Biology Society,* 2003, pp. 3613–3616.
32 Norton, G.S., Romlein, J., Lyche, D.K., "PACS 2000 – Quality Control using the Task Allocation Chart," *Proceedings of the SPIE Medical Imaging Conference,* 2000, Vol. 3980, pp. 384–389.
33 Okeyo, T. M., Adelhardt, M., Health Professionals Handbook on Quality Management in Healthcare in Kenya, Centre for Quality in Healthcare, Nairobi, Kenya, 2003.
34 Oliver, S., "The Healthy Standard," *Quality World,* Vol. 29, No. 2, 2003, pp. 16–20.
35 Owei, V., "Healthcare Quality and Productivity: Framework for an Information Technology Impact Architecture," *Proceedings of the Annual Meeting of the Decision Sciences Institute*, 1998, pp. 46–48.
36 Palmer, P., Mason, L., Dunn, M., "A Case Study in Healthcare Quality Management: A Practical Methodology for Auditing Total Patient X-ray Dose During a Diagnostic Procedure," *Proceedings of the 7th Biennial Conference on Engineering Systems Design and Analysis*, 2004, pp. 469–474.
37 Persson, J., Ekberg, K., Linden, M., "Work Organization, Work Environment and the Use of Medical Equipment: A Survey Study of the Impact on Quality and Safety," *Medical & Biological Engineering & Computing,* Vol. 31, No. 1, 1993, pp. 20–24.
38 Ransom, S. B., Joshi M., Nash, D., The Healthcare Quality Book: Vision, Strategy, and Tools, Health Administration Press, Chicago, Illinois, 2005.
39 Ray, P., Weerakkody, G., "Quality of Service Management in Healthcare Organizations: A Case Study," *Proceedings of the IEEE Symposium on Computer-Based Medical Systems,* 1999, pp. 80–85.
40 Reid, R.D., Christensen, M.M., "Quality Healthcare – A Path Forward," *Proceedings of the ASQ's 55th Annual Quality Congress,* 2001, pp. 57–63.
41 Revere, L., Black, K., Huq, A., "Integrating Six Sigma and CQI for Improving Patient Care," *TQM Magazine,* Vol. 16, No. 2, 2004, pp. 105–113.

42 Ribiere, V., La Salle, A.J., Khorramshahgol, R., "Hospital Information Systems quality: A Customer Satisfaction Assessment Tool," *Proceedings of the 32nd Annual Hawaii International Conference on System Sciences,* 1999, pp. 140.
43 Rooks, J., Zedick, J.L., "The Baldrige Criteria: Managing for Quality in Healthcare," *Proceedings of the 56th Annual Quality Congress*, 2002, pp. 549–550.
44 Sommer, T.J., "Telemedicine: A Useful and Necessary Tool to Improving Quality of Healthcare in the European Union," *Computer Methods and Programs in Biomedicine,* Vol. 48, No. 1–2, 1995, pp. 73–77.
45 Spicer, J., "How to Measure Patient Satisfaction," *Quality Progress,* Vol. 35, No. 2, 2002, pp. 97–98.
46 Stoecklein, M., "ASQ's Role in Healthcare," *Quality Progress,* Vol. 36, No. 3, 2003, pp. 90–91.
47 Tecca, M.B., Weitzner, W.M., "Measuring Performance in Home Health from Internal and External Vantage Points," *Proceedings of the 9th Annual Quest for Quality and Productivity in Health Services,* 1997, pp. 27–38.
48 Turk, A.R., Poulakos, E.M., "Practical Approaches for Healthcare: Indoor Air Quality Management," *Energy Engineering: Journal of the Association of Energy Engineering,* Vol. 93, No. 5, 1996, pp. 12–79.
49 Walters, D., Jones, P., "Value and Value Chains in Healthcare: A Quality Management Perspective," *TQM Magazine,* Vol. 13, No. 5, 2001, pp. 319–333.
50 Winterberg, L., "Quality Improvement in Healthcare," *Proceedings of the ASQ's 55th Annual Quality Congress,* 2001, pp. 352–353.

A.2.2 Internet Reliability

51 Amir, Y., Caudy, R., Munjal, A., Schlossnagle, T., Tutu, C., "N-Way Fail-Over Infrastructure for Reliable Servers and Routers," *Proceedings of the IEEE International Conference on Dependable Systems and Networks,* 2003, 130–135.
52 Barbeau, M., "Implementation of Two Approaches for the Reliable Multicast of Mobile Agents over Wireless Networks," *Proceedings of the International Symposium on Parallel Architectures, Algorithms and Networks, I-SPAN,* 1999, pp. 414–419.
53 Bell, S.J., Halperin, M., "Testing the Reliability of Cellular Online Searching," *Online (Wilton, Connecticut),* Vol. 19, No. 5, 1995, pp. 15–24.
54 Bonald, T., Massoulie, L., "Impact of Fairness on Internet Performance," *Proceedings of the Joint International Conference on Measurement and Modeling of Computer Systems,* Vol. 29, 2001, pp. 82–91.
55 Carlson, R., Hobby, R., Newman, H.B., "Measuring End-to-end Internet Performance," *Network Magazine,* Vol. 18, No. 4, 2003, pp. 42–46.
56 Chen, T.M., Oh, T.H., "Reliable Services in MPLS," *IEEE Communications Magazine,* Vol. 37, No. 12, 1999, pp. 58–62.
57 Choi, J., Lee, W., Kwon, D., "Quality Characteristics for the Test of Mobile Game Software," *Proceedings of the International Conference on Software Engineering Research and Practice,* 2004, pp. 564–570.
58 Chou, C., Peng, H., Hsieh, Y., "The Design and Development of an Internet Safety Curriculum in Taiwan," *Proceedings of the Seventh IASTED International Conference on Computers and Advanced Technology in Education,* 2004, pp. 546–551.

59 Chun Kin Chan, Tortorella, M., "Spares-inventory Sizing for End-to-end Service Availability," *Proceedings of the International Symposium on Product Quality and Integrity,* 2001, pp. 98–102.
60 Crovella, M., Lindemann, C., Reiser, M., "Internet Performance Modeling: The State of the Art at the Turn of the Century," *Performance Evaluation,* Vol. 42, No. 2, 2000, pp. 91–108.
61 Divan, D.M., Brumsickle, W.E., Luckiiff, G.A., "Real-time Power Quality and Reliability Monitoring for Utilities and their C and I Customers," *Proceedings of the Sixth International Conference on Advances in Power System Control, Operation and Management,* 2003, pp. 49–53.
62 Eguchi, T., Ohsaki, H., Murata, M., "Multivariate Analysis for Performance Evaluation of Active Queue Management Mechanisms in the Internet," *Internet Performance and Control of Network Systems,* 2002, pp. 144–153.
63 Elenien, A.R.A., Ismail, L.S., Bedor, H.S., "Quality of Service Handler for MPEG Video in Best Effort Environment," *Proceedings of the International Conference on Electrical, Electronic and Computer Engineering,* 2004, pp. 393–398.
64 Eslambolchi H., Danishmand M., "Reliability of Emerging Internet Based Services," Wiley, New York, 2003.
65 Field, J., Varela, C.A., "Transactors: A Programming Model for Maintaining Globally Consistent Distributed State in Unreliable Environments," *ACM SIGPLAN Notices,* Vol. 40, No. 1, 2005, pp. 195–208.
66 Fink, R.A., "Reliability Modeling of Freely-available Internet-distributed Software," *Proceedings of the International Software Metrics Symposium,* 1998, pp. 101–104.
67 Godrich, K.L., "Parameterization of Powering Solutions for Telecom/datacom Clients," *Proceedings of the 24th International Telecomunications Energy Conference,* 2002, pp. 273–278.
68 Goseva-Popstojanova, K., Mazimdar, S., Singh, A.D., "Empirical Study of Session-based Workload and Reliability for Web Servers," *Proceedings of the International Symposium on Software Reliability Engineering,* 2004, pp. 403–414.
69 Greco, R., "Satellite: Boosting Internet Performance," *Telecommunications (International Edition),* Vol. 31, No. 4, 1997, pp. 4–9.
70 Greenfield, D., "Storage, Heal Thyself!" *Network Magazine,* Vol. 19, No. 7, 2004, pp. 52–53.
71 Haungs, M., Pandey, R., Barr, E., "Handling Catastrophic Failures in Scalable Internet Applications," *Proceedings of the International Symposium on Applications and the Internet,* 2004, pp. 188–194.
72 Hecht, M., "Reliability/availability Modeling and Prediction for E-commerce and Other Internet Information Systems," *Proceedings of the International Symposium on Product Quality and Integrity,* 2001, pp. 176–182.
73 Hrasnica, H., Lehnert, R., "Performance Analysis of Error Handling Methods Applied to a Broadband PLC Access Network," *Proceedings of the Internet Performance and Control of Network Systems Conference,* Vol. 4865, 2002, pp. 166–177.
74 Jacko, J.A., Sears, A., and Sorensen, S.J., "Framework for Usability: Healthcare Professionals and the Internet," *Ergonomics,* Vol. 44, No. 11, 2001, pp. 989–1007.
75 Jauw, J., Vassiliou, P., "Field Data is Reliability Information: Implementing an Automated Data Acquisition and Analysis System," *Proceedings of the Annual Reliability and Maintainability Symposium,* 2000, pp. 86–93.

76 Kermarrec, A., Massoulie, L., and Ganesh, A.J., "Probabilistic Reliable Dissemination in Large-scale Systems," *IEEE Transactions on Parallel and Distributed Systems,* Vol. 14, No. 3, 2003, pp. 248–258.
77 Kim, D., Hong, W., Jong, M., "A Fault Management System for Reliable ADSL Services Provisioning," *Proceedings of the Internet Performance and Control of Network Systems Conference,* Vol. 4523, 2001, pp. 341–349.
78 Kogan Y., Choudhury, G., "Two Problems in Internet Reliability: New Questions for Old Models," *ACM Sigmetrics Performance Evaluation Review,* Vol. 32, No. 2, 2004, pp. 9–11.
79 Lee, S., "The Reliability Improvement of TCP for the Wireless Packet Data Network by Using the Virtual TCP Receive Window," *Proceedings of the IEEE Vehicular Technology Conference*, 2001, pp. 1866–1868.
80 Lee, S., Kim, J., Choi, J., "Development of a Web-based Integrity Evaluation System for Primary Components in a Nuclear Power Plant," *Proceedings of the Asian Pacific Conference on Nondestructive Testing,* 2004, pp. 2226–2231.
81 Levine, B.N., Lavo, D.B., Garcia-Luna-Aceves, J.J., "Case for Reliable Concurrent Multicasting Using Shared Ack Trees," *Proceedings of the ACM International Multimedia Conference,* 1996, pp. 365–376.
82 Lippy, B.E., "Technology Safety Data Sheets: The U.S. Department of Energy's Innovative Efforts to Mitigate or Eliminate Hazards During Design and to Inform Workers about the Risks of New Technologies," *Proceedings of the International Conference on Radioactive Waste Management and Environmental Remediation,* 2001, pp. 375–379.
83 Liu, H., Shooman, M.L., "Reliability Computation of an IP/ATM Network with Congestion," *Proceedings of the International Symposium on Product Quality and Integrity: Transforming Technologies for Reliability and Maintainability Engineering,* 2003, pp. 581–586.
84 Liu, Z., Almhana, J., and Choulakian, V., "Internet Performance Modeling using Mixture Dynamical System Models," *Proceedings of the International Conference on Pervasive Services,* 2004, pp. 189–198.
85 Lowry, E.S., "Software Simplicity and hence Safety – Thwarted for Decades," *Proceedings of the International Symposium on Technology and Society,* 2004, pp. 80–84.
86 Matthews, W., Cottrell, L., "PingER Project: Active Internet Performance Monitoring for the HENP Community," *IEEE Communications Magazine,* Vol. 38, No. 5, 2000, pp. 130–136.
87 McDowell, A., Schmidt, C., Yue, K., "Analysis and Metrics of XML Schema," *Proceedings of the International Conference on Software Engineering Research and Practice,* 2004, pp. 538–544.
88 Moh, M., Zhang, S., "Scalability Study of Application-level Reliable Multicast Congestion Control for the Next-generation Mobile Internet," *Proceedings of the International Conference on Third Generation Wireless and Beyond,* 2002, pp. 652–657.
89 Osman, T., Wagealla, W., Bargiela, A., "An Approach to Rollback Recovery of Collaborating Mobile Agents," *IEEE Transactions on Systems, Man and Cybernetics Part C: Applications and Reviews,* Vol. 34, No. 1, 2004, pp. 48–57.
90 Patel, R.B., Mastorakis, N., "Fault-tolerant Mobile Agents Computing," *WSEAS Transactions on Computers,* Vol. 4, No. 3, 2005, pp. 287–314.

91 Pawlicki, G., Sathaye, A., "Availability and Performance Oriented Availability Modeling of Webserver Cache Hierarchies," *Proceedings of the Annual Reliability and Maintainability Symposium: International Symposium on Product Quality and Integrity,* 2004, pp. 586–592.
92 Raman, S., McCanne, S., "Model Analysis and Protocol Framework for Soft State-based Communication," *Computer Communication Review,* Vol. 29, No. 4, 1999, pp. 15–25.
93 Rupe, J., "Performability Modeling and Decision Support for Computer Telephony Integration," *Proceedings of the International Symposium on Product Quality and Integrity: Transforming Technologies for Reliability and Maintainbility Engineering,* 2003, pp. 339–343.
94 Sahinoglu, M., Libby, D.L., "Sahinoglu-Libby (SL) Probability Density Function-component Reliability Applications in Integrated Networks," *Proceedings of the International Symposium on Product Quality and Integrity: Transforming Technologies for Reliability and Maintainbility Engineering,* 2003, pp. 280–287.
95 Salvador, P., Valadas, R., "A Framework Based on Markov Modulated Poisson Processes for Modeling Traffic with Long-range Dependence," *Proceedings of the Internet Performance and Control of Network Systems Conference,* 2001, pp. 221–232.
96 Saxena, A., Bhatia, S., "Identity Management for Improved Organisational Efficiency and E-business Growth: Managing Enterprise Knowledge," *International Journal of Information Technology and Management,* Vol. 4, No. 3, 2005, pp. 321–342.
97 Schwefel, H.-., Jobmann, M., Hollisch, D., "On the Accuracy of TCP Performance Models," *Proceedings of the Internet Performance and Control of Network Systems Conference,* 2001, pp. 91–102.
98 Sengupta, S., Oliaro, G., "The NSTX Trouble Reporting System," *Proceedings of the Symposium on Fusion Engineering,* 2002, pp. 242–244.
99 Shaw, R.D., Livingston, R.D., "Web-Based Occupational Health, Safety and Environmental (OHSE) Management Tools: Can They Help Companies Manage OHSE Performance in a Cost-Effective Manner?" *Proceedings of the SPE International Conference on Health, Safety and Environment in Oil and Gas Exploration and Production,* 2002, pp. 1551–1554.
100 Shelden, S., Vaughan, M., "The Internet Usability Engineer," *Ergonomics in Design,* Vol. 9, No. 2, 2001, pp. 27–28.
101 Shida, T., Yoshinari, Y., Hisatsune, M., "Development of Information Technology in the Construction and Maintenance of Nuclear Power Plants," *Hitachi Review,* Vol. 50, No. 3, 2001, pp. 73–78.
102 Stangel, M., Bharghavan, V., "Improving TCP Performance in Mobile Computing Environments," *Proceedings of the IEEE International Conference on Communications,* 1998, pp. 584–589.
103 Svitek, M., "Architecture of the Transport Telematics Applications Using the GNSS," *Proceedings of the International Conference on Information and Knowledge Engineering,* 2003, pp. 505–508.
104 Takagi, H., Kitajima, M., Yamamoto, T., "Search Process Evaluation for a Hierarchical Menu System by Markov Chains," *Proceedings of the Internet Performance and Control of Network Systems Conference,* 2001, pp. 183–192.

105 Takesue, T., "CG Technologies for Supporting Cooperative Creativity by Industrial Designers," *Proceedings of the IEEE International Workshop on Robot and Human Interactive Communication,* 2000, pp. 316–321.
106 To, M., Neusy, P., "Unavailability Analysis of Long-Haul Networks," IEEE Journal on Selected Areas in Communications, Vol. 12, No. 1, 1994, pp. 100–109.
107 Tse, P.W., He, L.S., "Web and Virtual Instrument Based Machine Remote Sensing, Monitoring and Fault Diagnostic System," *Proceedings of the Biennial Conference on Mechanical Vibration and Noise,* 2001, pp. 2919–2926.
108 Veitch, P., "A Survivable and Cost-Effective IP Metro Interconnect Architecture," *IEEE Communications Magazine,* Vol. 41, No. 12, 2003, pp. 100–105.
109 Waite, D.A., "Internet Information Sources – Quantity Vs. Quality," *Proceedings of the ASQ's 52nd Annual Quality Congress,* 1998, pp. 344–348.
110 Wang, G., Cao, J., Chan, K.C.C., "RGB: A Scalable and Reliable Group Membership Protocol in Mobile Internet," *Proceedings of the International Conference on Parallel Processing,* 2004, pp. 326–333.
111 Wishart, J., "Internet Safety in Emerging Educational Contexts," *Computers and Education,* Vol. 43, No. 1–2, 2004, pp. 193–204.
112 Wunnava, S.V., Jasani, H., "Secure Multimedia Activity with Redundant Schemes," *Proceedings of the IEEE Southeast Conference,* 2002, pp. 333–337.
113 Yamaguchi, M., Yamamoto, M., "Congestion Control Scheme for Reliable Multicast Improving Intrasession Fairness with Network Support," *Electronics and Communications in Japan, Part I: Communications (English Translation of Denshi Tsushin Gakkai Ronbunshi),* Vol. 88, No. 2, 2005, pp. 61–70.
114 Yang, S., Chou, H., "Adaptive QoS Parameters Approach to Modeling Internet Performance," *International Journal of Network Management,* Vol. 13, No. 1, 2003, pp. 69–82.
115 Yao, B., Fuchs, W.K., "Proxy-based Recovery for Applications on Wireless Hand-held Devices," *Proceedings of the IEEE Symposium on Reliable Distributed Systems,* 2000, pp. 2–10.
116 Yoon, W., Lee, D., Youn, H.Y., "A Combined Group/tree Approach for Scalable Many-to-many Reliable Multicast," *IEEE Infocom,* Vol. 3, 2002, pp. 1336–1345.
117 Yoshimura, M., Hamaguchi, N., Kozaki, T., "High-performance and Data-Optimized 3rd-generation Mobile Communication System: 1xEV-DO," *Hitachi Review,* Vol. 53, No. 6, 2004, pp. 271–275.

A.2.3 Quality Control in the Food Industry

118 Al-Habaibeh, A., Shi, F., Brown, N., "A Novel Approach for Quality Control System Using Sensor Fusion of Infrared and Visual Image Processing for Laser Sealing of Food Containers," *Measurement Science and Technology,* Vol. 15, No. 10, 2004, pp. 1995–2000.
119 Alli, I., Food Quality Assurance: Principles and Practices, CRC Press, Boca Raton, Florida , 2004.
120 Ashpole, C., "Quality Control in Automated Food Processing," *Quality Assurance (Engineer),* Vol. 8, No. 3, 1982, pp. 81–85.
121 Avramescu, A., Andreescu, S., Noguer, T., "Biosensors Designed for Environmental and Food Quality Control Based on Screen-printed Graphite Electrodes with Different Configurations," *Analytical and Bioanalytical Chemistry,* Vol. 374, No. 1, 2002, pp. 25–32.

122 Baldwin, E.K., "Online Near Infrared Spectroscopy for Measurement, Control, and Quality Assurance in the Food Processing Industry," *Proceedings of the Conference on Food Processing Automation*, 1992, pp. 254–257.
123 Balfoort, A.J., "Food Quality and Farm Quality: A Difference by Nature," *Proceedings of the 26th EOQC Conference, Amsterdam*, 1982, pp. 10–14.
124 Barni, M., Mussa, A.W., Mecocci, A., "Intelligent Perception System for Food Quality Inspection Using Color Analysis," *Proceedings of the IEEE International Conference on Image Processing. Part 1 (of 3), Vol. 1,* 1995, pp. 450–453.
125 Bednarjevsky, S.S., Veryasov, Y.V., Akinina, E.V., "Modern Methods and Systems of the Precise Control of the Quality of the Agricultural and Food Production," *Proceedings of SPIE Conference,* Vol. 3543, 1999, pp. 376–384.
126 Belton, P.S., "Spectroscopic Approaches to the Measurement of Food Quality," *Pure and Applied Chemistry,* Vol. 69, No. 1, 1997, pp. 47–50.
127 Birth, G.S., Eisler, G., "Optics for Food Quality Analysis," *Proceedings of the SPIE Conference on Opt. in Quality Assurance*, Vol. 170, 1979, pp. 1979.
128 Brera, C., Miraglia, M., "Proficiency Testing Programmes as a Tool in Food Quality Assurance: Overview of Italian Experiences Materials in the Life Sciences," *Mikrochimica Acta,* Vol. 123, No. 1–4, 1996, pp. 39–45.
129 Bro, R., Van den Berg, F., Thybo, A., "Multivariate Data Analysis as a Tool in Advanced Quality Monitoring in the Food Production Chain," *Trends in Food Science and Technology,* Vol. 13, No. 6–7, 2002, pp. 235–244.
130 Brosnan, T., Sun, D., "Improving Quality Inspection of Food Products by Computer Vision – A Review," *Journal of Food Engineering,* Vol. 61, No. 1, 2004, pp. 3–16.
131 Castillo, O., Melin, P., "Automated Quality Control in the Food Industry Combining Artificial Intelligence Techniques with Fractal Theory," *Proceedings of the 10th International Conference on Applications of Artificial Intelligence in Engineering,* 1995, pp. 109–118.
132 Castillo, O., Melin, P., "Intelligent System for the Identification of Microorganisms for Quality Control in the Food Industry," *Proceedings of the 9th International Conference on Applications of Artificial Intelligence in Engineering,* 1994, pp. 133–140.
133 Castoldi, F., "HACCP: From Quality Control to Quality Assurance: Comparison of Strategies to the Advantage of Food Industry Operators," *Industrie Alimentari,* Vol. 35, No. 345, 1996, pp. 121–125.
134 Costa, A.I.A., Dekker, M., Jongen, W.M.F., "Quality Function Deployment in the Food Industry: A Review," *Trends in Food Science and Technology,* Vol. 11, No. 9–10, 2000, pp. 306–314.
135 Crew, S.M., "Quality Assurance and Food Safety: An Enforcement Officer's Perspective," *Quality Forum,* Vol. 17, No. 4, 1991, pp. 165–170.
136 Crozier, L.L., "Quality-related Education in Food Processing," *Proceedings of the 46th Annual ASQC Quality Congress Transactions* , Vol. 46, 1992, pp. 734–740.
137 Cys, R.J., Surak, J.G., "Designing Quality into a Food Plant," *Proceedings of the 45th Annual ASQC Quality Congress Transactions,* Vol. 45, 1991, pp. 31–35.
138 Davies, R., Heleno, P., Correia, B., "VIP3D – An Application of Image Processing Technology for Quality Control in the Food Industry," *Proceedings of the IEEE International Conference on Image Processing,* 2001, pp. 293–296.
139 Davis, M., "Spotlight – An Individual's Perspective on Quality in the Food Industry," *Quality World,* 1998, pp. 14–16.

140 Deshpande, S.S., Rocco, R.M., "Biosensors and their Potential Use in Food Quality Control," *Food Technology,* Vol. 48, No. 6, 1994, pp. 146–150.
141 Douglas, D.E., Weatherly, S.H., "Total Quality Control System for Food Production in Hospitals," *Proceedings of the 32nd Annual Conference of the American Society for Quality Control,* 1978, pp. 566–573.
142 Dris, R., Sharma, A., Food Technology and Quality Evaluation, Science Publishers, Enfield, New Hampshire, 2003.
143 Druce, E., "Practical Basis of Quality Assurance in Food Manufacture," *Proceedings of the World Quality Congress*, 1984, pp. 417–428.
144 Du, C., Sun, D., "Recent Developments in the Applications of Image Processing Techniques for Food Quality Evaluation," *Trends in Food Science and Technology,* Vol. 15, No. 5, 2004, pp. 230–249.
145 Farkas, J., Mohacsi-Farkas, C., "Application of Differential Scanning Calorimetry in Food Research and Food Quality Assurance," *Journal of Thermal Analysis,* Vol. 47, No. 6, 1996, pp. 1787–1803.
146 Forsgren, G., Frisell, H., Ericsson, B., "Taint and Odour Related Quality Monitoring of Two Food Packaging Board Products using Gas Chromatography, Gas Sensors and Sensory Analysis," *Nordic Pulp and Paper Research Journal,* Vol. 14, No. 1, 1999, pp. 5–16.
147 Funazaki, N., Hemmi, A., Ito, S., "Application of Semiconductor Gas Sensor to Quality Control of Meat Freshness in Food Industry," *Sensors and Actuators, B: Chemical,* Vol. B25, No. 1–3 pt 2, 1995, pp. 797–800.
148 Gaonkar, A., McPherson, A., *Ingredient Interactions: Effects on Food Quality*, Taylor & Francis, Boca Raton, Florida, 2006.
149 Gates, K.W., "Automation for Food Engineering: Food Quality Quantization and Process Control," *Journal of Aquatic Food Product Technology,* Vol. 11, No. 3–4, 2002, pp. 317–322.
150 Giese, J.H., "Emerging Food Quality Assessment Techniques," *Food Technology,* Vol. 55, No. 12, 2001, pp. 68–70.
151 Goldsworthy, B., "Systematic Approach to Quality Control in the Food Industry," *Quality Assurance (Engineer),* Vol. 4, No. 4, 1978, pp. 111–115.
152 Golomski, W.A., "Total Quality Management and the Food Industry. Why is it Important?" *Food Technology,* Vol. 47, No. 5, 1993, pp. 74–78.
153 Goodall, P.G., "British Food Industry and the Single Market: Quality Issues in Perspective," *Quality Forum,* Vol. 16, No. 3, 1990, pp. 143–148.
154 Goyache, F., Bahamonde, A., Alonso, J., "The Usefulness of Artificial Intelligence Techniques to Assess Subjective Quality of Products in the Food Industry," *Trends in Food Science and Technology,* Vol. 12, No. 10, 2001, pp. 370–381.
155 Gunasekaran, S., Ding, K., "Using Computer Vision for Food Quality Evaluation," *Food Technology,* Vol. 48, No. 6, 1994, pp. 151–154.
156 Horowitz, J.K., "Regulating Safety and Quality Standards in Food Marketing, Processing, and Distribution: Discussion," *American Journal of Agricultural Economics,* Vol. 78, No. 5, 1996, pp. 1261–1264.
157 Huang, Y., Lacey, R.E., Whittaker, A.D., "Food Quality Quantization: Theory and Practice," *Proceedings of the ASAE Annual International Meeting,* 2000, pp. 1073–1094.
158 Jones, C., "Food Quality Reference Materials," *Food Science and Technology,* Vol. 18, No. 3, 2004, pp. 17–21.

159 Kerr, D., Shi, F., Brown, N., "Quality Inspection of Food Packaging Seals using Machine Vision with Texture Analysis," *Proceedings of the Institution of Mechanical Engineers, Part B: Journal of Engineering Manufacture,* Vol. 218, No. 11, 2004, pp. 1591–1599.
160 Kim, M.S., Chen, Y.R., Mehl, P.M., "Hyperspectral Reflectance and Fluorescence Imaging System for Food Quality and Safety," *Transactions of the American Society of Agricultural Engineers,* Vol. 44, No. 3, 2001, pp. 721–729.
161 Kollar, G., Syposs, Z., Viczian, G., "Quality Management System as a Tool of Process Control for Food and Agro Industries," *Hungarian Journal of Industrial Chemistry,* Vol. 29, No. 2, 2001, pp. 135–138.
162 Kseibat, D., Basir, O.A., Mittal, G.S., "Artificial Neural Network for Optimizing Safety and Quality in Thermal Food Processing," *Proceedings of the IEEE International Symposium on Intelligent Control,* 1999, pp. 393–398.
163 Locht, P., Thomsen, K., Mikkelsen, P., "Full Color Image Analysis as a Tool for Quality Control and Process Development in the Food Industry," *Proceedings of the ASAE Annual International Meeting*, 1997, pp. 15–20.
164 MacArthur, M.D., "The Food Quality Protection Act Replaces the Delaney Clause with a Negligible Risk Standard for Pesticide Residues, Possibly Signaling More Reforms for Food Safety Standards," *Paper, Film and Foil Converter,* Vol. 70, No. 11, 1996, pp. 22–25.
165 McLaughlin, W.L., NBS, W., DC, Miller, A., "Radiation Dosimetry for Quality Control of Food Preservation and Disinfestation," *Transactions of the 4th International Meeting on Radiation Processing,* Vol. 22, 1983, pp. 21–29.
166 Molnar, P., "Quality Trends in the Food Industry of Hungary," *Proceedings of the World Quality Congress*, 1984, pp. 369–374.
167 Najjar, L.J., Thompson, J.C., Ockerman, J.J., "Wearable Computer for Quality Assurance Inspectors in a Food Processing Plant," *Proceedings of the 1st International Symposium on Wearable Computers,* 1997, pp. 163–164.
168 Ni, H., Gunasekaran, S., "Food Quality Prediction with Neural Networks," *Food Technology,* Vol. 52, No. 10, 1998, pp. 60–66.
169 Nilsson, H., Tuncer, B., Thidell, A., "The Use of Eco-labeling like Initiatives on Food Products to Promote Quality Assurance – Is There Enough Credibility?" *Journal of Cleaner Production,* Vol. 12, No. 5, 2004, pp. 515–524.
170 O'Farrell, M., Lewis, E., Flanagan, C., "An Intelligent Optical Fibre Based Sensor System for Monitoring Food Quality," *Proceedings of the First IEEE International Conference on Sensors,* 2002, pp. 308–311.
171 O'Farrell, M., Lewis, E., Flanagan, C., "Combining Principal Component Analysis with an Artificial Neural Network to Perform Online Quality Assessment of Food as it Cooks in a Large-scale Industrial Oven," *Sensors and Actuators, B: Chemical,* Vol. 107, No. 1, 2005, pp. 104–112.
172 O'Farrell, M., Lewis, E., Flanagan, C., "Controlling a Large-Scale Industrial Oven by Monitoring the Food Quality, both Internally and Externally, Using an Optical Fibre Based System," *Proceedings of the IEEE Second International Conference on Sensors*, 2003, pp. 368–371.
173 O'Farrell, M., Lewis, E., Flanagan, C., "Intelligent Processing of Spectroscopic Signals Obtained Using an Optical Fibre Based System for Food Quality Control," *International Journal of Smart Engineering System Design,* Vol. 5, No. 4, 2003, pp. 409–416.

174 O'Farrell, M., Lewis, E., Flanagan, C., "Monitoring Food Quality Utilising an Intelligent Optical Fiber Based Sensor System," *Proceedings of the Artificial Neural Networks in Engineering Conference*, 2002, pp. 957–962.
175 O'Farrell, M., Lyons, W.B., Lewis, E., "A Method for Assessing Quality of Food Products Based on Optical Spectra Using Intelligent Pattern Recognition in a Full Scale Production Environment," *Proceedings of the Artificial Neural Networks in Engineering Conference*, 2003, pp. 945–950.
176 Ong, K.G., Puckett, L.G., Sharma, B.V., "Wireless, Passive, Resonant-circuit Sensors for Monitoring Food Quality," *Proceedings of the Chemical and Biological Early Warning Monitoring for Water, Food, and Ground Conference*, 2001, pp. 150–159.
177 Pearce, S.J., "Food Processing Quality System Guideline RE ANSI/ASQC Z1. 15," *Proceedings of the 40th Anniversary Quality Congress Transactions*, 1986, pp. 557–566.
178 Pei, J., Chen, T., "Food Quality of the Formosan Sika Deer in the Cheting area, Kenting, Southern Taiwan," *Taiwan Journal of Forest Science*, Vol. 19, No. 4, 2004, pp. 353–362.
179 Perrot, N., Bonazzi, C., Trystram, G., "Estimation of the Food Product Quality Using Fuzzy Sets," *Proceedings of the Annual Conference of the North American Fuzzy Information Processing Society*, 1999, pp. 487–491.
180 Plichta, J.C., "Total Quality Management in the Fast-food Industry," *Food Technology*, Vol. 48, No. 9, 1994, pp. 152.
181 Porter, A.M., "Quality Trends in the Food Industry," *Proceedings of the World Quality Congress*, 1984, pp. 429–436.
182 Puri, S.C., "Food Safety and Quality Control: SPC with HACCP," *Proceedings of the Forty-Fourth Annual Quality Congress Transactions*, 1990, pp. 729–735.
183 Roudot, A., "Terminology in Food Quality Management," *Proceedings of the 5th International Conference on Quality, Reliability, and Maintenance*, 2004, pp. 39–42.
184 Schröder, M. J. A., Food Quality and Consumer Value: Delivering Food That Satisfies, Springer, New York, 2003.
185 Scotter, C.N.G., "Non-destructive Spectroscopic Techniques for the Measurement of Food Quality," *Trends in Food Science & Technology*, Vol. 8, No. 9, 1997, pp. 285–292.
186 Sestak, J., Zitny, R., Houska, M., "Simple Rheological Models of Food Liquids for Process design and Quality Assessment," *Journal of Food Engineering*, Vol. 1, No. 1, 1983, pp. 35–49.
187 Shaw, S., Rose, S.A., "New Food Legislation and the Role of Quality Assurance," *Quality Forum*, Vol. 17, No. 4, 1991, pp. 151–157.
188 Shchedrina, T.M., and Vorotenitskaya, S.L., "Assessment of Quality in Food Industry and Motivation of its Improvement," *Proceedings of the World Quality Congress*, 1984, pp. 356–360.
189 Shewfelt, R.L., Erickson, M.C., Hung, Y., "Applying Quality Concepts in Frozen Food Development," *Food Technology*, Vol. 51, No. 2, 1997, pp. 56–59.
190 Shi, F., Brown, N., Kerr, D., "A Machine Vision Approach to Quality Inspection of Laser Welded, Semi-rigid Food Containers," *Proceedings of the International Conference on Mechatronics*, 2003, pp. 383–388.

191 Sjoeberg, A.M., "Quality Systems for Trouble-shooting in Food Processes," *Proceedings of the Conference on New Shelf-Life Technologies and Safety Assessments,* 1995, pp. 183–188.
192 Stringer, D., "Quality Circles in the Food Industry," *Proceedings of the 26th EOQC Conference Amsterdam*, 1982, pp. 4–6.
193 Sun, D., "Computer vision – An Objective, Rapid and Non-contact Quality Evaluation Tool for the Food Industry," *Journal of Food Engineering,* Vol. 61, No. 1 SPEC, 2004, pp. 1–2.
194 Sun, Y., Zhao, X., Tan, J., "Automatic Measuring of Apparent Quality of Extruded Food," *Nongye Jixie Xuebao/Transactions of the Chinese Society of Agricultural Machinery,* Vol. 30, No. 1, 1999, pp. 63–67.
195 Sun, Y., Zhao, X., Tan, J., "Intelligent Control Methods for Extruded Food Quality," *Nongye Gongcheng Xuebao/Transactions of the Chinese Society of Agricultural Engineering,* Vol. 14, No. 1, 1998, pp. 183–187.
196 Surak, J.G., "ISO 9000: Part of a Quality System for Food Companies," *Proceedings of the 47th Annual Quality Congress,* 1993, pp. 781–788.
197 Szalai, G., "Use of PCR Techniques in Food Quality Control," *Periodica Polytechnica, Chemical Engineering,* Vol. 40, No. 1–2, 1996, pp. 101–104.
198 Tolboe, O., "Quality Trends in the Food Industry," *Proceedings of the World Quality Congress*, 1984, pp. 409–416.
199 Valls, E.R., "New Trends in Food Quality Control," *Proceedings of the World Quality Congress*, 1984, pp. 69–71.
200 Van Der Spiegel, M., Luning, P.A., Ziggers, G.W., "Towards a Conceptual Model to Measure Effectiveness of Food Quality Systems," *Trends in Food Science and Technology,* Vol. 14, No. 10, 2003, pp. 424–431.
201 Whitworth, P., "Approach to Quality in the Food Industry," *Proceedings of the World Quality Congress*, 1984, pp. 375–382.
202 Winter, C.K., "EPA's implementation of the Food Quality Protection Act," *Food Technology,* Vol. 52, No. 11, 1998, pp. 148–151.
203 Ziggers, G.W., Trienekens, J., "Quality Assurance in Food and Agribusiness Supply Chains: Developing Successful Partnerships," *International Journal of Production Economics,* Vol. 60–61, 1999, pp. 271–279.

A.2.4 Quality Control in the Textile Industry

204 Abed, M.H., Dale, B.G., "Attempt to Identify Quality-related Costs in Textile Manufacturing," *Quality Assurance,* Vol. 13, No. 2, 1987, pp. 41–45.
205 Acaccia, G.M., Marelli, A., Michelini, R.C., "Automatic Fabric Storing and Feeding in Quality Clothing Manufacture," *Journal of Intelligent and Robotic Systems: Theory and Applications,* Vol. 37, No. 4, 2003, pp. 443–465.
206 Anagnostopoulos, C., Anagnostopoulos, I., Vergados, D., "High Performance Computing Algorithms for Textile Quality Control," *Mathematics and Computers in Simulation,* Vol. 60, No. 3–5, 2002, pp. 389–400.
207 Anon, "Process Control: Making it Happen," *Textile World*, Vol. 132, No. 11, 1982, pp. 58, 63–65.
208 Anonymous "Impact of BS 5750 and Total Quality Management on the Textile Industry," *Quality World,* Vol. 20, No. 7, 1994, pp. 454.

209 Aschner, G.S., "Quality Control in the textile Industry," *Quality Assurance (Engineer)*, Vol. 6, No. 1, 1980, pp. 19–21.
210 Aschner, G.S., "Some Aspects of the Product Development and Quality," *Proceedings of the ASQC 40th Quality Congress Transactions, Milwaukee*, WI, USA, 1986, pp. 479–483.
211 Aschner, G.S., Koczy, L.R., Salman, A.A., "Quality Planning in One Area of the Textile Industry: What Can be Produced from a Given Raw Material?", *Proceedings of the International Conference on Quality Control*, 1978, pp. 355–360.
212 Benisek, L., "Serious TQM," *Textile Month*, No. CT, 2001, pp. 46–47.
213 Bottcher, H.H., Rieder, O., "Industrial Textile Making-up Important Quality Assurance Parameters," *International Textile Bulletin: Nonwovens, Industrial Textiles*, Vol. 47, No. 4, 2001, pp. 34–36.
214 Carr, C.M., Roberts, J.C., "Technology Transfer. A Quality Control Tool from the Textile Industry," *Paper Technology*, Vol. 34, No. 9, 1993, pp. 27–28.
215 Dallmann, H., "Measuring Technology and Quality Control," *International Textile Bulletin*, Vol. 49, No. 6, 2003, pp. 58–60.
216 Dallmann, H., "Testing Technology and Quality Control," *International Textile Bulletin*, Vol. 45, No. 4, 1999, pp. 118–119.
217 Darby, A.D.J., Gowdy, J.N., Wolla, M.L., "Quality Control of Textile Color by Minicomputer/Microcomputer," *Proceedings of the IEEE SOUTHEAST Conference*, 1979, pp. 329–332.
218 Djiev, S.N., Pavlov, L.I., "Low-cost Automated System for On-line Quality and Production Control in Textiles," *Process Control and Quality*, Vol. 7, No. 3–4, 1995, pp. 179–183.
219 El Mogahzy, Y.E., "Using Off-line Quality Engineering in Textile Processing, Part I: Concepts and Theories," *Textile Research Journal*, Vol. 62, No. 5, 1992, pp. 266–274.
220 Elbert, H., Fuss, F., Heald, T., "Test Methods for the Quality Assurance of Dyes," *Textile Chemist and Colorist*, Vol. 20, No. 9, 1988, pp. 63–67.
221 Erenyi, I., Pongracz, J., "Quality Control in Textile Industry via Machine Vision," *Microprocessing and Microprogramming*, Vol. 32, No. 1–5, 1991, pp. 807–813.
222 Esswein, R., "Knowledge Assures Quality," *International Textile Bulletin*, Vol. 50, No. 2, 2004, pp. 17–20.
223 Gopalakrishnan P., "Applications of Statistical Techniques in Textile Industry," *Proceedings of the International Conference on Quality Control*, 1969, pp. 583–586.
224 Grover, E.B., Handbook of Textile Testing and Quality Control, Textile Book Publishers, New York, 1960.
225 Hanada, T., Guide Book on Quality Control in Textile Industry, Asian Productivity Organization, Tokyo, 1967.
226 Hawkyard, C.J., McShane, I., "Quest for Quality in the British Textile Industry," *Journal of the Textile Institute*, Vol. 85, No. 4, 1994, pp. 469–475.
227 Heleno, P., Davies, R., Brazio Correia, B.A., "A Machine Vision Quality Control System for Industrial Acrylic Fibre Production," *Eurasip Journal on Applied Signal Processing*, No. 7, 2002, pp. 728–735.
228 Huart, J., Postaire, J.G., "Integration of Computer Vision onto Weavers for Quality Control in the Textile Industry," *Proceedings of the SPIE Machine Vision Applications in Industrial Inspection II Conference*, 1994, Vol. 2183, pp. 155–163.

229 Karkanis, S., Tsoutsou, K., Metaxaki-Kossionidis, C., "An On-line Quality Inspection System for Textile Industries," *Proceedings of the IEEE 6th Mediterranean Electrotechnica Conferencel,* 1991, pp. 1255–1259.
230 Kassaee, M., "Taking the Strategic Approach Towards Quality. An Exploratory Field Study of the Textile Industry," *Proceedings of the Forty-Fourth Annual Quality Congress Transactions,* 1990, Vol. 44, pp. 121–127.
231 Konev, D.G., "Measurement and Data System of Man-made Fiber Quality Control," *Fiber Chemistry,* Vol. 11, No. 1, 1979, pp. 72–76.
232 Lees, G., "Quality Philosophies and Practices in the Garment Business," *Quality Assurance,* Vol. 11, No. 1, 1985, pp. 21–24.
233 Mahall, K., Quality Assessment of Textiles: Damage Detection by Microscopy, Springer, New York, 2003.
234 Meier, R., Uhlmann, J., Leuenberger, R., "Uster Fabriscan Automatic Quality Inspection System for Fabrics," *Melliand Textilberichte/International Textile Reports,* Vol. 80, No. 5, 1999, pp. 96–97.
235 Mueller, H., "New Technology in Quality Control and On-line Quality Assurance in the Textile Industry in the 80's," *Proceedings of the Symposium on New Technologies for Cotton,* Port Elizabeth, South Africa, 1982, pp. 220–238.
236 Ratnam, T. V., Chellamani, K.P., *Quality Control in Spinning, 3rd rev. edn*., South India Textile Research Association, Coimbatore, India,1999.
237 Rupp, J., "Albini: Obsessed by Quality," *International Textile Bulletin,* Vol. 50, No. 5, 2004, pp. 51–54.
238 Rupp, J., "Customer Service Knows No Bounds," *International Textile Bulletin,* Vol. 47, No. 3, 2001, pp. 6–16.
239 Rupp, J., Bohringer, A., "Quality Comes Before Price," *International Textile Bulletin,* Vol. 44, No. 2, 1998, pp. 10–11.
240 Sahin, U.K., Gursoy, N.C., "Low Temperature Acidic Pectinase Scouring for Enhancing Textile Quality," *AATCC Review,* Vol. 5, No. 1, 2005, pp. 27–30.
241 Salusso-Deonier, C.J., "Gaining a Competitive Edge with Top Quality Sizing," *Proceedings of the 43rd Annual Quality Congress Transactions*, 1989, Vol. 3, pp. 371–376.
242 Sari-Sarraf, H., Goddard, J.S., "Vision System for On-loom Fabric Inspection," *Proceedings of the IEEE Annual Textile, Fiber and Film Industry Technical Conference,* 1998, pp. 11–15.
243 Seidel, L.E., "Texturing QC: The Problems of Growth," *Textile Industries,* Vol. 141, No. 7, 1977, pp. 25–28.
244 Sigmon, D.M., Grady, P.L., Winchester, S.C., "Computer Integrated Manufacturing and Total Quality Management," Textile Progress, Vol. 27, 1998, pp. 1–56.
245 Stevens, R.B., "QC in Woven Upholstery," *Textile Industries,* Vol. 139, No. 7, 1975, pp. 47–50.
246 Suh, M.W., "Quality Process, and Cost Controls. A 'Random Walk' in Textile Profitability," *Journal of the Textile Institute,* Vol. 83, No. 3, 1992, pp. 348–360.
247 Tantaswadi, P., Vilainatre, J., Tamaree, N., "Machine Vision for Automated Visual Inspection of Cotton Quality in Textile Industries Using Color Isodiscrimination Contour," *Computers and Industrial Engineering,* Vol. 37, No. 1–2, 1999, pp. 347–350.
248 Thomas, T., Cattoen, M., "Automatic Inspection of Simply Patterned Material in the Textile Industry," *Proceedings of the SPIE Machine Vision Applications in Industrial Inspection Conference,* 1994, Vol. 2183, pp. 2–12.

249 Topf, W., "Quality Management for the Textile Industry," *Journal of Coated Fabrics,* Vol. 25, 1996, pp. 285–300.
250 Turbet, C., "Use of Gas-infrared Hoods for Improved Quality and Productivity in the Paper and Textile Industries," *Proceedings of the SPE Gas Technology Symposium,* 1989, pp. 97–102.
251 Whiteley, K.J., "Mather Lecture Quality Control in the Processing of Wool and the Performance of Wool Textiles," *Journal of the Textile Institute,* Vol. 79, No. 3, 1988, pp. 339–348.
252 Winchester, S.C., "Total Quality Management in Textiles," *Journal of the Textile Institute,* Vol. 85, No. 4, 1994, pp. 445–459.
253 Winchester, S.C., Sigmon, D.L., Grady, P.L., "Integrating Total Quality Management and Computer Integrated Manufacturing in Textiles," *Proceedings of the 77th World Conference of the Textile Institute. Part 1 (of 2), Vol. 1*, 1997, pp. 155–178.
254 Wulfhorst, B., "Evaluation of the New Spinning Techniques on Automation and Quality Criteria," *International Textile Bulletin,* Vol. 36, No. 4, 1990, pp. 5–10.
255 Wulfhorst, B., Schaepers, J., "Quality and Economy in Cotton Yarn Production," *Textile Month*, 1984, pp. 56–59.

A.2.5 Software Quality

256 Al-Janabi, A., Aspinwall, E., "Using Quality Design Metrics to Monitor Software Development," *Quality World,* 1996, pp. 25–34.
257 Anjard, R., "Software Quality Assurance Considerations," *Microelectronics and Reliability,* Vol. 35, No. 6, 1995, pp. 995–1000.
258 "Software, Design and Quality Control," *Modern Plastics,* Vol. 78, No. 10, 2001, pp. 34–38.
259 "Improving Product Quality and Spotting Potential Problems with Software," *Electrical Design and Manufacturing,* Vol. 9, No. 4, 1995, pp. 13–14.
260 "Total Quality Software," *Process Engineering,* Vol. 73, No. 6, 1992, pp. 43–46.
261 April, A., Abran, A., Dumke, R.R., "SMCMM Model to Evaluate and Improve the Quality of the Software Maintenance Process," *Proceedings of the European Conference on Software Maintainance and Reengineering,* 2004, pp. 243–248.
262 Ares, J., Pazos, J., "Conceptual Modelling: An Essential Pillar for Quality Software Development," *Knowledge-Based Systems,* Vol. 11, No. 2, 1998, pp. 87–104.
263 Arnoult, W.S., "Quality Software Development Through Effective Project Management," *Proceedings of the Project Management Institute Annual Seminar/Symposium,* 1991, pp. 135–138.
264 Asada, M., Yan, P.M., "Strengthening Software Quality Assurance," *Hewlett-Packard Journal,* Vol. 49, No. 2, 1998, pp. 89–97.
265 Ashrafi, N., "Decision Making Framework for Software Total Quality Management," *International Journal of Technology Management,* Vol. 16, No. 4–6, 1998, pp. 532–543.
266 Azuma, M., Komiyama, T., Miyake, T., "Panel: The Model and Metrics for Software Quality Evaluation Report of the Japanese National Working Group," *Proceedings of the 14th Annual International Computer Software and Applications Conference*, 1990, pp. 64–69.

267 Baisch, E., Liedtke, T., "Comparison of Conventional Approaches and Soft-computing Approaches for Software Quality Prediction," *Proceedings of the IEEE International Conference on Systems, Man, and Cybernetics,* 1997, pp. 1045–1049.
268 Baker, R.A.J., "Code Reviews Enhance Software Quality," *Proceedings of the IEEE 19th International Conference on Software Engineering,* 1997, pp. 570–571.
269 Barrett, T., "Dancing with Devils: Or Facing the Music on Software Quality," *Electronic Design,* Vol. 45, No. 13, 1997, pp. 117–120.
270 Beaver, J.M., Schiavone, G.A., "A Comparison of Software Quality Modeling Techniques," *Proceedings of the International Conference on Software Engineering Research and Practice*, 2003, pp. 263–266.
271 Bennett, P., "Experience in Engineering Quality into Software," *Software Engineering Journal,* Vol. 11, No. 2, 1996, pp. 95–98.
272 Bevan, N., Azuma, M., "Quality in Use: Incorporating Human Factors into the Software Engineering Lifecycle," *Proceedings of the 3rd IEEE International Software Engineering Standards Symposium and Forum,* 1997, pp. 169–179.
273 Binder, L.H., Poore, J.H., "Field Experiments with Local Software Quality Metrics," *Software – Practice and Experience,* Vol. 20, No. 7, 1990, pp. 631–647.
274 Bishop, D.C., Pymms, P., "Software Quality Management," *Nuclear Engineer: Journal of the Institution of Nuclear Engineers,* Vol. 32, No. 4, 1991, pp. 128–131.
275 Blaha, M., "A Copper Bullet for Software Quality Improvement," *Computer,* Vol. 37, No. 2, 2004, pp. 21–25.
276 Boegh, J., Depanfilis, S., and Kitchenham, B., "Method for Software Quality Planning, Control, and Evaluation," *IEEE Software,* Vol. 16, No. 2, 1999, pp. 69–77.
277 Bolstad, M., "Design by Contract: A Simple Technique for Improving the Quality of Software," *Proceedings of the DoD HPCMP Users Group Conference,* 2004, pp. 303–307.
278 Boone, R., Lucas, K., Wynd, R., "Practical Quality Metrics for Resolution Enhancement Software," *Proceedings of the SPIE Cost and Performance in Integrated Circuit Creation Conference,* Vol. 5043, 2003, pp. 162–171.
279 Bouktif, S., Kegl, B., Sahraoui, H., "Combining Software Quality Predictive Models: An Evolutionary Approach," *Proceedings of the IEEE International Conference on Software Maintenance,* 2002, pp. 385–392.
280 Brandt, W.D., Wilson, D.N., "Measure of Software Quality," *Proceedings of the 2nd International Conference on Software Quality Management*, 1994, pp. 73–75.
281 Bukowski, J.V., Goble, W.M., "Practical Lessons for Improving Software Quality," *Proceedings of the Annual Reliability and Maintainability Symposium,* 1990, pp. 436–440.
282 Bunse, C., Verlage, M., Giese, P., "Improved Software Quality through Improved Development Process Descriptions," *Automatica,* Vol. 34, No. 1, 1998, pp. 23–32.
283 Burns, P.J., "Software metrics. The Key to Quality Software on the NCC project," *Proceedings of the Networks Technology Conference,* 1993, pp. 27–30.
284 Burr, A., "Quality Software – A Success Story?" *Quality World,* Vol. 27, No. 3, 2001, pp. 4–5.
285 Burton, S., Swanson, K., Leonard, L., "Quality and Knowledge in Software Engineering," *AI Magazine,* Vol. 14, No. 4, 1993, pp. 43–50.
286 Bush, M., "Improving Software Quality: The Use of Formal Inspections at the Jet Propulsion Laboratory," *Proceedings of the 12th International Conference on Software Engineering,* 1990, pp. 196–199.

287 Cangussu, J.W., Mathur, A.P., Karcich, R.M., "Software Release Control Using Defect Based Quality Estimation," *Proceedings of the 15th International Symposium on Software Reliability Engineering,* 2004, pp. 440–450.
288 Card, D.N., "Managing Software Quality with Defects," *Proceedings of the 26th Annual International Computer Software and Applications Conference,* 2002, pp. 472–474.
289 Card, D.N., "Software Quality Engineering," *Information and Software Technology,* Vol. 32, No. 1, 1990, pp. 1–10.
290 Carvallo, J.P., Franch, X., Grau, G., "QM: A Tool for Building Software Quality Models," *Proceedings of the 12th IEEE International Requirements Engineering Conference,* 2004, pp. 358–359.
291 Cechich, A., Piattini M., Vallecillo, A., *Component-based Software Quality: Methods and Techniques*, Springer, New York, 2003.
292 Chen, D.J., Chen, W.C., Huang, S.K., "Survey of the Influence of Programming Constructs and Mechanisms on Software Quality," *Journal of Information Science and Engineering,* Vol. 10, No. 2, 1994, pp. 177–201.
293 Cho, J., Lee, S.J., "An Evaluation Model for Software Quality Improvement," *Proceedings of the International Conference on Software Engineering Research and Practise*, 2003, pp. 615–620.
294 Choi, J., Park, S., Chong, K., "ISO/IEC 9126 Quality Characteristics Considering the Application Fields and the Test Phases of Embedded Software," *Proceedings of the International Conference on Software Engineering Research and Practice,* 2004, pp. 628–632.
295 Cobb, R.H., Mills, H.D., "Engineering Software under Statistical Quality Control," *IEEE Software,* Vol. 7, No. 6, 1990, pp. 45–54.
296 Cosgriff, P.S., "Quality Assurance of Medical Software," *Journal of Medical Engineering & Technology,* Vol. 18, No. 1, 1994, pp. 1–10.
297 Cosgriff, P.S., Vauramo, E., Pretschner, P., "QAMS. Quality Assurance of Medical Software," *Proceedings of the Advances in Medical Informatics: Results of the AIM Exploratory Action Conference,* 1992, pp. 310–313.
298 Daughtrey, T., Editor, *Fundamental Concepts for the Software Quality Engineer*, ASQ Quality Press, Milwaukee, Wisconsin, 2002.
299 Davis, C.J., Thompson, J.B., Smith, P., "Current Approaches to Software Quality Assurance within the United Kingdom," *Proceedings of the International Conference on Software Quality Management,* 1993, pp. 849–854.
300 Delcambre, S.N., Rainey, V.P., Tanik, M.M., "Defining Quality to Support the Software Engineering Activity," *Proceedings of the Energy-Sources Technology Conference and Exhibition,* 1995, pp. 93–99.
301 Demirors, O., "Assumptions and Difficulties of Software Quality Movement," *Proceedings of the 23rd EUROMICRO Conference,* 1997, pp. 115–122.
302 Dette, W., "Software Product Assurance IBM'S Software Quality Assurance Function," *Forensic Engineering,* Vol. 2, No. 1–2, 1990, pp. 89–96.
303 Dickson, J., "The RQMS as an Evolving System of Software Quality Measurements," *Proceedings of the International Conference on Communications*, 1991, pp. 1743–1746.
304 Dolle, P., Jackson, K., "Experiences of Software Quality Management in Software Maintenance – A Case Study," *Proceedings of the International Conference on Software Quality Management,* 1993, pp. 629–634.

305 Drake, T., "Measuring Software Quality: A Case Study," *Computer,* Vol. 29, No. 11, 1996, pp. 78–87.
306 Dromey, R.G., "Model for Software Product Quality," *IEEE Transactions on Software Engineering,* Vol. 21, No. 2, 1995, pp. 146–162.
307 Dromey, R.G., Bailes, C., Xiaoge, L., "Model for Enhancing Software Product Quality," *Proceedings of the 16th Australian Computer Science Conference*, 1993, pp. 461–465.
308 Dubrovin, V.I., Doroshenko, Y.N., "Software Quality Estimation," *Upravlyayushchie Sistemy i Mashiny,* No. 5, 2001, pp. 34–39.
309 Dugan, J.B., Sullivan, K.J., Coppit, D., "Developing a High-quality Software Tool for Fault Tree Analysis," *Proceedings of the International Symposium on Software Reliability Engineering,* 1999, pp. 222–231.
310 Dumke, R.R., Kuhran, I., "Tool-based Quality Management in Object-oriented Software Development," *Proceedings of the 3rd Symposium on Assessment of Quality Software Development Tools,* 1994, pp. 148–160.
311 Duncan, S.P., Martin, C.R., Quigley-Lawrence, R., " 'Customers' and 'Users': Two Faces of Software Quality and Productivity," *Proceedings of the IEEE International Conference on Communications*, 1990, pp. 15–18.
312 Dyer, M., "Cleanroom Approach to Quality Software Development," *Proceedings of the CMG '92 Conference*, 1992, pp. 1201–1212.
313 Ebert, C., Morschel, I., "Environment for Measuring and Improving the Quality of Object-Oriented Software," *Quality and Reliability Engineering International,* Vol. 15, No. 1, 1999, pp. 33–45.
314 Ebert, C., Morschel, I., "Metrics for Quality Analysis and Improvement of Object-oriented Software," *Information and Software Technology,* Vol. 39, No. 7, 1997, pp. 497–509.
315 Eickelmann, N., Hayes, J.H., "New Year's Resolutions for Software Quality," *IEEE Software,* Vol. 21, No. 1, 2004, pp. 12–13.
316 Elboushi, M.I., Sherif, J.S., "Object-oriented Software Design Utilizing Quality Function Deployment," *Journal of Systems and Software,* Vol. 38, No. 2, 1997, pp. 133–143.
317 Erikkson, I., McFadden, F., "Quality Function Deployment: A Tool to Improve Software Quality," *Information and Software Technology,* Vol. 35, No. 9, 1993, pp. 491–498.
318 Esaki, K., Yamada, S., Takahashi, M., "A Quality Engineering Approach to Human Factors Affecting Software Reliability in Design Process," *Electronics and Communications in Japan,* Vol. 85, No. 3, 2002, pp. 33–42.
319 Evans, I., *Achieving Software Quality through Teamwork*, Artech House, Boston, Massachusetts, 2004.
320 Faith, B.J., "Making a Better Job of Software Quality with Total Quality Management," *Proceedings of the International Conference on Software Quality Management,* 1993, pp. 73–76.
321 Fallah, M.H., Jrad, A.M., "SQA – A Proactive Approach to Assuring Software Quality," *AT&T Technical Journal,* Vol. 73, No. 1, 1994, pp. 26–33.
322 Farbey, B., "Software Quality Metrics: Considerations about Requirements and Requirement Specifications," *Information and Software Technology,* Vol. 32, No. 1, 1990, pp. 60–64.

323 Feng, J., Tang, R., Wang, S., "Study on Software Quality Grey Quantitative Evaluation Mode," *Harbin Gongye Daxue Xuebao/Journal of Harbin Institute of Technology*, Vol. 37, No. 5, 2005, pp. 639–641.
324 Finnie, B.W., Johnston, I.H.A., "Acceptance of Software Quality as a Contributor to Safety," *Proceedings of the Safety and Reliability Society Symposium*, 1990, pp. 279–284.
325 Fishburn, S., "Integration of a Total Quality Management Program through Software Aided Design, Qualification, Planning, and Scheduling Tools," *Proceedings of the International SAVE Conference*, 1992, pp. 233–239.
326 Frey, G., "Software Quality in Logic Controller Programming," *Proceedings of the IEEE International Conference on Systems, Man and Cybernetics*, 2002, pp. 516–521.
327 Galdi, V., Ippolito, L., Piccolo, A., "On Industrial Software Quality Assessment," *Proceedings of the 3rd International Conference on Software for Electrical Engineering Analysis and Design*, 1996, pp. 501–509.
328 Galin, D., *Software Quality Assurance*, Pearson Education Limited, New York, 2004.
329 Ganesan, K., Khoshgoftaar, T.M., Allen, E.B., "Case-based Software Quality Prediction," *International Journal of Software Engineering and Knowledge Engineering*, Vol. 10, No. 2, 2000, pp. 139–152.
330 Ghods, M., Nelson, K.M., "Contributors to Quality during Software Maintenance," *Decision Support Systems*, Vol. 23, No. 4, 1998, pp. 361–369.
331 Gong, B., Yen, D.C., Chou, D.C., "Manager's Guide to Total Quality Software Design," *Industrial Management and Data Systems*, Vol. 98, No. 3–4, 1998, pp. 100–107.
332 Gonzalez, R.R., "Unified Metric of Software Complexity: Measuring Productivity, Quality, and Value," *Journal of Systems and Software*, Vol. 29, No. 1, 1995, pp. 17–37.
333 Gorton, I., Chan, T.S., Jelly, I., "Engineering High Quality Parallel Software using PARSE," *Proceedings of the 3rd Joint International Conference on Vector and Parallel Processing*, 1994, pp. 381.
334 Guglielmi, N., Guerrieri, R., "Experimental Comparison of Software Methodologies for Image Based Quality Control," *Proceedings of the 20th International Conference on Industrial Electronics, Control and Instrumentation*, 1994, pp. 1942–1945.
335 Gulezian, R., "Software Quality Measurement and Modeling, Maturity, Control and Improvement," *Proceedings of the 2nd IEEE International Software Engineering Standards Symposium*, 1995, pp. 52–59.
336 Gyorkos, J., Rozman, I., Vajde Horvat, R., "Quality Management in Software Development Process: An Empirical Model," *Proceedings of the 1996 IEEE International Engineering Management Conference*, 1996, pp. 191–195.
337 Haag, S., Raja, M.K., Schkade, L.L., "Quality Function Deployment Usage in Software Development," *Communications of the ACM*, Vol. 39, No. 1, 1996, pp. 41–49.
338 Hadfield, P., Hope, S., "Survey of Classification Schemes for Improving Software Quality and An Outline for their Future," *Proceedings of the International Conference on Software Quality Management*, 1993, pp. 613–617.

339 Haitani, L.N., "Applying Manufacturing Quality Control Theory to Performance-oriented Software Development," *Proceedings of the CMG '92 Conference*, 1992, pp. 890–894.
340 Hall, T., Wilson, D., "Views of Software Quality: A Field Report," *Proceedings of the IEE Software Conference,* Vol. 144, No. 2, 1997, pp. 111–118.
341 Hanna, M.S., Loew-Blosser, W., "SQEngineer: A Methodology and Tool for Specifying and Engineering Software Quality," *Proceedings of the Symposium on Assessment of Quality Software Development Tools,* 1992, pp. 194–210.
342 Haque, T., "PCM – The Smart Way to Achieve Software Quality," *Quality World,* Vol. 23, No. 11, 1997, pp. 922–923.
343 Haugh, J.M., "Never Make the Same Mistake Twice – Using Configuration Control and Error Analysis to Improve Software Quality," *Proceedings of the IEEE/AIAA 10th Digital Avionics Systems Conference,* 1991, pp. 220–225.
344 He, Z., Staples, G., Ross, M., "Fourteen Japanese Quality Tools in Software Process Improvement," *TQM Magazine,* Vol. 8, No. 4, 1996, pp. 40–44.
345 Henry, S., and Goff, R., "Comparison of a Graphical and a Textual Design Language Using Software Quality Metrics," *Journal of Systems and Software,* Vol. 14, No. 3, 1991, pp. 133–146.
346 Herbsleb, J., Zubrow, D., Goldenson, D., "Software Quality and the Capability Maturity Model," *Communications of the ACM,* Vol. 40, No. 6, 1997, pp. 30–40.
347 Herron, D., "Early Life Cycle Identification of Software Quality Risk Factors," *Proceedings of the ASQ's 52nd Annual Quality Congress,* 1998, pp. 399–400.
348 Hevner, A.R., "Phase Containment Metrics for Software Quality Improvement," *Information and Software Technology,* Vol. 39, No. 13, 1997, pp. 867–877.
349 Hindel, B., "How to Ensure Software Quality for Real Time Systems," *Control Engineering Practice,* Vol. 1, No. 1, 1993, pp. 35–41.
350 Hirayama, M., Sato, H., Yamada, A., "Practice of Quality Modeling and Measurement on Software Life-cycle," *Proceedings of the 12th International Conference on Software Engineering,* 1990, pp. 98–107.
351 Hoffmann, H., "How to Improve Overall System Quality by a Sensible Design of Man-machine Interaction: Views of a Software Engineer," *Proceedings of the European Safety & Reliability Conference,* 1993, pp. 939–944.
352 Hon, S.E.I., "Assuring Software Quality through Measurements. A Buyer's Perspective," *Journal of Systems and Software,* Vol. 13, No. 2, 1990, pp. 117–130.
353 Hong, G.Y., Goh, T.N., "Six Sigma in Software Quality," *TQM Magazine,* Vol. 15, No. 6, 2003, pp. 364–373.
354 Hopker, O., "Pragmatic Approach to Software Quality for Information Systems in Small and Medium Sized Organization," *Proceedings of the 3rd International Conference on Software Quality Management,* 1995, pp. 75–80.
355 Horch, J.W., Practical Guide to Software Quality Management, 2nd ed., Artech House, Boston, Massachusetts, 2003.
356 Horgan, J.R., London, S., Lyu, M.R., "Achieving Software Quality with Testing Coverage Measures," *Computer,* Vol. 27, No. 9, 1994, pp. 60–70.
357 Houston, D., Keats, J.B., "Cost of Software Quality: A Means of Promoting Software Process Improvement," *Quality Engineering,* Vol. 10, No. 3, 1998, pp. 563–567.
358 Hun Oh, S., Yeon Lee, J., Lee, Y.J., "Software Quality Manager: A Knowledge-based Management Tool of Software Metrics," *Proceedings of the 1994 IEEE Region 10's 9th Annual International Conference (TENCON'94)*, 1995, pp. 796–800.

359 Huo, M., Verner, J., Zhu, L., "Software Quality and Agile Methods," *Proceedings of the 28th Annual International Computer Software and Applications Conference*, 2004, pp. 520–525.
360 Jack, R., "Customer Involvement in the Application of Quality Management Systems to Software Product Development," *Proceedings of the Computing and Control Division Colloquium on Customer Driven Quality in Product Design*, 1994, pp. 6.1–6.5.
361 Jeffery, R., Basili, V., Berry, M., "Establishing Measurement for Software Quality Improvement," *Proceedings of the IFIP TC8 Open Conference on Business Process Re-engineering: Information Systems Opportunities and Challenges*, 1994, pp. 319–329.
362 Johnson, C.S., Wilson, D.N., "Using Quality Principles to Teach Software Quality Techniques," *International Journal of Environmental Studies A & B*, Vol. 47, No. 1, 1995, pp. 465–468.
363 Johnson, P.M., "Instrumented Approach to Improving Software Quality through Formal Technical Review," *Proceedings of the 16th International Conference on Software Engineering*, 1994, pp. 113–122.
364 Jones, D.R., Murthy, V., Blanchard, J., "Quality and Reliability Assessment of Hardware and Software during the Total Product Life Cycle," *Quality and Reliability Engineering International*, Vol. 8, No. 5, 1992, pp. 477–483.
365 Jones, W.D., Hudepohl, J.P., Khoshgoftaar, T.M., "Application of a Usage Profile in Software Quality Models," *Proceedings of the 3rd European Conference on Software Maintenance and Reengineering*, 1999, pp. 148–157.
366 Jorgensen, M., "Measurement of Software Quality," *Proceedings of the 1st International Conference on Software Quality Engineering*, 1997, pp. 257–266.
367 Jorgensen, M., "Software Quality Measurement," *Advances in Engineering Software*, Vol. 30, No. 12, 1999, pp. 907–912.
368 Joshi, S.M., Misra, K.B., "Quantitative Analysis of Software Quality during the Design and Implementation Phase," *Microelectronics and Reliability*, Vol. 31, No. 5, 1991, pp. 879–884.
369 Juliff, P., "Software Quality Function Deployment," *Proceedings of the 2nd International Conference on Software Quality Management*, 1994, pp. 533–536.
370 Jung, H., Choi, B., "Optimization Models for Quality and Cost of Modular Software Systems," *European Journal of Operational Research*, Vol. 112, No. 3, 1999, pp. 613–619.
371 Jung, H., Kim, S., Chung, C., "Measuring Software Product Quality: A Survey of ISO/IEC 9126," *IEEE Software*, Vol. 21, No. 5, 2004, pp. 88–92.
372 Kan, S.H., Basili, V.R., Shapiro, L.N., "Software Quality: An Overview from the Perspective of Total Quality Management," *IBM Systems Journal*, Vol. 33, No. 1, 1994, pp. 4–19.
373 Kan, S.H., Dull, S.D., Amundson, D.N., "AS/400 Software Quality Management," *IBM Systems Journal*, Vol. 33, No. 1, 1994, pp. 62–88.
374 Kan, S.H., Metrics and Models in Software Quality Engineering, Addison-Wesley, Boston, Massachusetts, 2003.
375 Kaplan, C., "Secrets of Software Quality at IBM," *Proceedings of the ASQC's 50th Annual Quality Congress*, 1996, pp. 479–484.
376 Kaufer, S., Stacy, W., "Software Quality Begins with a Quality Design," *Object Magazine*, Vol. 5, No. 8, 1996, pp. 56–60.

377 Kautz, K., Ramzan, F., "Software Quality Management and Software Process Improvement in Denmark," *Proceedings of the 34th Annual Hawaii International Conference on System Sciences,* 2001, pp. 126–129.
378 Keene, S.J.J., "Cost Effective Software Quality," *Proceedings of the Annual Reliability and Maintainability Symposium,* 1991, pp. 433–437.
379 Kelly, D.P., Oshana, R.S., "Improving Software Quality using Statistical Testing Techniques," *Information and Software Technology,* Vol. 42, No. 12, 2000, pp. 801–807.
380 Kemp, A., "Software Quality Giving Managers What They Need," *Engineering Management Journal,* Vol. 5, No. 5, 1995, pp. 235–238.
381 Kenett, R.S., "Software Specifications Metrics: A Quantitative Approach to Assess the Quality of Documents," *Proceedings of the 19th Convention of Electrical and Electronics Engineers in Israel,* 1996, pp. 166–169.
382 Khaled, E.E., The ROI from Software Quality, Auerbach Publications, Boca Raton, Florida, 2005.
383 Khoshgoftaar, T.M., Allen, E.B., "Impact of Costs of Misclassification on Software Quality Modeling," *Proceedings of the 4th International Software Metrics Symposium,* 1997, pp. 54–62.
384 Khoshgoftaar, T.M., Allen, E.B., "Practical Classification-rule for Software-quality Models," *IEEE Transactions on Reliability,* Vol. 49, No. 2, 2000, pp. 209–216.
385 Khoshgoftaar, T.M., Allen, E.B., Deng, J., "Controlling Overfitting in Software Quality Models: Experiments with Regression Trees and Classification," *Proceedings of the 7th International Software Metrics Symposium*, 2001, pp. 190–198.
386 Khoshgoftaar, T.M., Allen, E.B., Halstead, R., "Using Process History to Predict Software Quality," *Computer,* Vol. 31, No. 4, 1998, pp. 66–72.
387 Khoshgoftaar, T.M., Allen, E.B., Kalaichelvan, K.S., "Predictive Modeling of Software Quality for Very Large Telecommunications Systems," *Proceedings of the IEEE International Conference on Communications,* 1996, pp. 214–219.
388 Khoshgoftaar, T.M., Halstead, R., Allen, E.B., "Process Measures for Predicting Software Quality," *Proceedings of the High-Assurance Systems Engineering Workshop,* 1997, pp. 155–160.
389 Khoshgoftaar, T.M., Lanning, D.L., "On the Impact of Software Product Dissimilarity on Software Quality Models," *Proceedings of the 4th International Symposium on Software Reliability Engineering,* 1994, pp. 104–114.
390 Khoshgoftaar, T.M., Munson, J.C., "Software Metrics and the Quality of Telecommunication Software," *Proceedings of Tricomm: High-Speed Communications Networks,* 1992, pp. 255–260.
391 Khoshgoftaar, T.M., Munson, J.C., Bhattacharya, B.B., "Predictive Modeling Techniques of Software Quality from Software Measures," *IEEE Transactions on Software Engineering,* Vol. 18, No. 11, 1992, pp. 979–987.
392 Khoshgoftaar, T.M., Seliya, N., "Comparative Assessment of Software Quality Classification Techniques: An Empirical Case Study," *Empirical Software Engineering,* Vol. 9, No. 3, 2004, pp. 229–257.
393 Khoshgoftaar, T.M., Seliya, N., "Fault Prediction Modeling for Software Quality Estimation: Comparing Commonly Used Techniques," *Empirical Software Engineering,* Vol. 8, No. 3, 2003, pp. 255–283.
394 Khoshgoftaar, T.M., Seliya, N., Herzberg, A., "Resource-oriented Software Quality Classification Models," *Journal of Systems and Software,* Vol. 76, No. 2, 2005, pp. 111–126.

395 Khoshgoftaar, T.M., Shan, R., Allen, E.B., "Improving Tree-based Models of Software Quality with Principal Components Analysis," *Proceedings of the 11th International Symposium on Software Reliability Engineering*, 2000, pp. 198–209.
396 Khoshgoftaar, T.M., Szabo, R.M., Guasti, P.J., "Exploring the Behaviour of Neural Network Software Quality Models," *Software Engineering Journal,* Vol. 10, No. 3, 1995, pp. 89–96.
397 Khoshgoftaar, T.M., Yuan, X., Allen, E.B., "Balancing Misclassification Rates in Classification-tree Models of Software Quality," *Empirical Software Engineering,* Vol. 5, No. 4, 2000, pp. 313–330.
398 Kikuchi, N., Mizuno, O., Kikuno, T., "Identifying Key Attributes of Projects that Affect the Field Quality of Communication Software," *Proceedings of the IEEE 24th Annual International Computer Software and Applications Conference,* 2000, pp. 176–178.
399 Kitchenham, B., Pfleeger, S.L., "Software Quality: The Elusive Target," *IEEE Software,* Vol. 13, No. 1, 1996, pp. 12–21.
400 Knox, S.T., "Modeling the Cost of Software Quality," *Digital Technical Journal,* Vol. 5, No. 4, 1993, pp. 9–12.
401 Koethluk Fllorescu, A., "EURESCOM project P227. Quality Assurance of Software for Telecommunication Systems," *Proceedings of the 3rd International Conference on Software Quality Management,* 1995, pp. 51–55.
402 Kokol, P., Chrysostalis, M., Bogonikolos, N., "Software Quality founded on Design Laws," *Proceedings of the Seventh IASTED International Conference on Software Engineering and Applications,* 2003, pp. 728–732.
403 Krasner, H., "Exploring the Principles that Underlie Software Quality Costs," Proceedings of the ASQ *Annual Quality Congress,* 1999, pp. 500–503.
404 Lac, C., Raffy, J., "Tool for Software Quality," *Proceedings of the Symposium on Assessment of Quality Software Development Tools,* 1992, pp. 144–150.
405 Laitenberger, O., "Studying the Effects of Code Inspection and Structural Testing on Software Quality," *Proceedings of the 9th International Symposium on Software Reliability Engineering,* 1998, pp. 237–246.
406 Lanubile, F., Visaggio, G., "Evaluating Predictive Quality Models Derived from Software Measures: Lessons Learned," *Journal of Systems and Software,* Vol. 38, No. 3, 1997, pp. 225–234.
407 Lauesen, S., Younessi, H., "Is Software Quality Visible in the Code?", *IEEE Software,* Vol. 15, No. 4, 1998, pp. 69–73.
408 Le Gall, G., "Software Quality Control," *Commutation & Transmission,* Vol. 17, No. 3, 1995, pp. 91–101.
409 Lee, M., Shih, K., Huang, T., "Well-evaluated Cohesion Metrics for Software Quality," *Ruan Jian Xue Bao/Journal of Software,* Vol. 12, No. 10, 2001, pp. 1447–1463.
410 Leffingwell, D.A., Norman, B., "Software Quality in Medical Devices: A Top-down Approach," *Proceedings of the 6th Annual IEEE Symposium on Computer-Based Medical Systems,* 1993, pp. 307–31.
411 Lehner, F., "Quality Control in Software Documentation: Measurement of Text Comprehensibility," *Information & Management,* Vol. 25, No. 3, 1993, pp. 133–137.
412 Lem, E., Software Metrics: The Discipline of Software Quality, Book Surge, North Charleston, South Carolina, 2005.

413 Lewis, N.D.C., "Assessing the Evidence from the Use of SPC in Monitoring, Predicting & Improving Software Quality," *Computers and Industrial Engineering,* Vol. 37, No. 1–2, 1999, pp. 157–160.

414 Linberg, K.R., "Defining the Role of Software Quality Assurance in a Medical Device Company," *Proceedings of the 6th Annual IEEE Symposium on Computer-Based Medical Systems,* 1993, pp. 278–283.

415 Lindermeier, R., "Quality Assessment of Software Prototypes," *Reliability Engineering & System Safety,* Vol. 43, No. 1, 1994, pp. 87–94.

416 Liu, X.F., "Quantitative Approach for Assessing the Priorities of Software Quality Requirements," *Journal of Engineering and Applied Science,* Vol. 42, No. 2, 1998, pp. 105–113.

417 Lloyd, I.J., Simpson, M.J., "Legal Aspects of Software Quality," *International Conference on Software Quality Management,* 1993, pp. 247–252.

418 Lounis, H., Ait-Mehedine, L., "Machine-learning Techniques for Software Product Quality Assessment," *Proceedings of the Fourth International Conference on Quality Software,* 2004, pp. 102–109.

419 Lowe, J., Daughtrey, T., Jensen, B., "Software Quality. International Perspectives," *Proceedings of the 47th Annual Quality Congress,* 1993, pp. 893–894.

420 Lowe, J.E., Jensen, B., "Customer Service Approach to Software Quality," *Proceedings of the 46th Annual Quality Congress,* 1992, pp. 1077–1083.

421 Luo, X., Zhan, J., Mao, M., "New Model of Improving Quality Management in China Software Company," *Journal of Computational Information Systems,* Vol. 1, No. 1, 2005, pp. 175–177.

422 Mallory, S.R., "Building Quality into Medical Product Software Design," *Biomedical Instrumentation & Technology,* Vol. 27, No. 2, 1993, pp. 117–135.

423 Mandeville, W.A., "Software Costs of Quality," *IEEE Journal on Selected Areas in Communications,* Vol. 8, No. 2, 1990, pp. 315–318.

424 Mantyla, M.V., "Developing New Approaches for Software Design Quality Improvement Based on Subjective Evaluations," *Proceedings of the 26th International Conference on Software Engineering,* 2004, pp. 48–50.

425 Mao, M., Luo, X., "Software Quality Management and Software Process Model," *Journal of Information and Computation Science,* Vol. 1, No. 3, 2004, pp. 203–207.

426 Marriott, P.C., McCorkhill, B.S., Lamb, R.A., "Software Quality Networking for Professionals," *Proceedings of the Forty-Fourth Annual Quality Congress Transactions,* 1990, pp. 511–517.

427 Maselko, W.T., Winstead, L.S., Kahn, R.E., "Streamlined Software Development Process: A Proven Solution to the Cost/schedule/quality challenge," *Proceedings of the Annual Forum of the American Helicopter Society,* Vol. 1, 1999, pp. 12–35.

428 McDonough, J.A., "Template for Software Quality Management for Department of Defense Programs," *Proceedings of the IEEE National Aerospace and Electronics Conference*, 1990, pp. 1281–1283.

429 McParland, P., "Using Function Point Analysis to Help Assess the Quality of a Software System," *Proceedings of the 3rd International Conference on Software Quality Management,* 1995, pp. 127–130.

430 Meskens, N., "Software Quality Analysis Systems: A New Approach," *Proceedings of the IEEE 22nd International Conference on Industrial Electronics, Control, and Instrumentation,* 1996, pp. 1406–1411.

431 Mirel, B., Olsen, L.A., Prakash, A., "Improving Quality in Software Engineering through Emphasis on Communication," *Proceedings of the ASEE Annual Conference,* 1997, pp. 9–12.
432 Mohamed, W.E.A., Siakas, K.V., "Assessing Software Quality Management Maturity (SQMM). A new model incorporating technical as well as cultural factors," *Proceedings of the 3rd International Conference on Software Quality Management,* 1995, pp. 325–328.
433 Moore, B.J., "Achieving Software Quality through Requirements Analysis," *Proceedings of the IEEE International Engineering Management Conference,* 1994, pp. 78–83.
434 Moores, T.T., Champion, R.E.M., "Software Quality through the Traceability of Requirements Specifications," *Proceedings of the 1st International Conference on Software Testing, Reliability and Quality Assurance,* 1994, pp. 100–104.
435 Moses, J., "Software Quality Methodologies," *International Conference on Software Quality Management,* 1993, pp. 333–337.
436 Moxey, C., "Experiences with Quality on a Recent Software Development Project," *Proceedings of the International Conference on Software Quality Management,* 1993, pp. 375–378.
437 Munson, J.C., and Khoshgoftaar, T.M., "Regression Modelling of Software Quality: Empirical Investigation," *Information and Software Technology,* Vol. 32, No. 2, 1990, pp. 106–114.
438 Murine, G.E., "Using the Rome Laboratory Framework and Implementation Guidebook as the Basis for an International Software Quality Metric Standard," *Proceedings of the 2nd IEEE International Software Engineering Standards Symposium,* 1995, pp. 61–70.
439 Murugesan, S., "Attitude Towards Testing: A Key Contributor to Software Quality," *Proceedings of the 1st International Conference on Software Testing, Reliability and Quality Assurance,* 1994, pp. 111–115.
440 Nagarajan, S.V., Garcia, O., Croll, P., "Software Quality Issues in Extreme Programming," *Proceedings of the 21st IASTED International Multi-Conference on Applied Informatics,* 2003, pp. 1090–1095.
441 Nance, R.E., Managing Software Quality: A Measurement Framework for Assessment and Prediction, Springer, New York, 2002.
442 Nathan, P., System Testing with an Attitude: An Approach that Nurtures Front-loaded Software Quality, Dorset House Pub., New York, 2005.
443 Neal, M.L., "Managing Software Quality through Defect Trend Analysis," *Proceedings of the PMI Annual Seminar/Symposium,* 1991, pp. 119–122.
444 Offutt, J., "Quality Attributes of Web Software Applications," *IEEE Software,* Vol. 19, No. 2, 2002, pp. 25–32.
445 Ogasawara, H., Yamada, A., Kojo, M., "Experiences of Software Quality Management using Metrics through the Life-cycle," *Proceedings of the 18th International Conference on Software Engineering,* 1996, pp. 179–188.
446 Olagunju, A.O., "Concepts of Operational Software Quality Metrics," *Proceedings of the 20th Annual ACM Computer Science Conference*, 1992, pp. 301–308.
447 Opiyo, E.Z., Horvath, I., Vergeest, J.S.M., "Quality Assurance of Design Support Software: Review and Analysis of the State of the Art," *Computers in Industry,* Vol. 49, No. 2, 2002, pp. 195–215.
448 O'Regan, G., A Practical Approach to Software Quality, Springer, New York, 2002.

449 Osmundson, J.S., Michael, J.B., Machniak, M.J., "Quality Management Metrics for Software Development," *Information and Management,* Vol. 40, No. 8, 2003, pp. 799–812.
450 Parnas, D.L., Lawford, M., "Inspection's Role in Software Quality Assurance," *IEEE Software,* Vol. 20, No. 4, 2003, pp. 16–20.
451 Parnas, D.L., Lawford, M., "The Role of Inspection in Software Quality Assurance," *IEEE Transactions on Software Engineering,* Vol. 29, No. 8, 2003, pp. 674–676.
452 Parzinger, M.J., Nath, R., "Effects of TQM Implementation Factors on Software Quality," *Proceedings of the Annual Meeting of the Decision Sciences Institute,* 1997, pp. 834–836.
453 Paulish, D.J., "Methods and Metrics for Developing High Quality Patient Monitoring System Software," *Proceedings of the Third Annual IEEE Symposium on Computer-Based Medical Systems Conference,* 1990, pp. 145–152.
454 Pedrycz, W., Peters, J.F., Ramanna, S., "Software Quality Measurement: Concepts and Fuzzy Neural Relational Model," *Proceedings of the IEEE International Conference on Fuzzy Systems,* 1998, pp. 1026–1031.
455 Pelnik, T.M., Suddarth, G.J., "Implementing Training Programs for Software Quality Assurance Engineers," *Medical Device and Diagnostic Industry,* Vol. 20, No. 10, 1998, pp. 75–80.
456 Pence, J.L., Hon, S.E., "Software Surveillance – A Buyer Quality Assurance Program," *IEEE Journal on Selected Areas in Communications,* Vol. 8, No. 2, 1990, pp. 301–308.
457 Peslak, A.R., "Improving Software Quality: An Ethics Based Approach," *Proceedings of the 2004 ACM SIGMIS CPR Conference,* 2004, pp. 144–149.
458 Pfleeger, S.L., "Software Quality," *Dr.Dobb's Journal of Software Tools for Professional Programmer,* Vol. 23, No. 3, 1998, pp. 22–27.
459 Phan, D.D., George, J.F., Vogel, D.R., "Managing Software Quality in a Very Large Development Project," *Information & Management,* Vol. 29, No. 5, 1995, pp. 277–283.
460 Pierce, K.M., "Why have a Software Quality Assurance Program?" *Nuclear Plant Journal,* Vol. 12, No. 5, 1994, pp. 57–61.
461 Pivka, M., "Software Quality System in a Small Software House," *Proceedings of the 3rd International Conference on Software Quality Management,* 1995, pp. 83–86.
462 Plant, R.T., "Factors in Software Quality for Knowledge-based Systems," *Information and Software Technology,* Vol. 33, No. 7, 1991, pp. 527–536.
463 Post, D.E., Kendall, R.P., "Software Project Management and Quality Engineering Practices for Complex, Coupled Multiphysics, Massively Parallel Computational Simulations: Lessons learned from ASCI," *International Journal of High Performance Computing Applications,* Vol. 18, No. 4, 2004, pp. 399–416.
464 Pratt, W.M., "Experiences in the Application of Customer-based Metrics in Improving Software Service Quality," *Proceedings of the International Conference on Communications*, 1991, pp. 1459–1462.
465 Putnik, Z., "On the Quality of the Software Produced," *Proceedings of the International Conference on Computer Science and Informatics,* 1998, pp. 64–66.
466 Rai, A., Song, H., Troutt, M., "Software Quality Assurance: An Analytical Survey and Research Prioritization," *Journal of Systems and Software,* Vol. 40, No. 1, 1998, pp. 67–83.

467 Ramakrishnan, S., "Quality Factors for Resource Allocation Problems – Linking Domain Analysis and Object-oriented Software Engineering," *Proceedings of the 1st International Conference on Software Testing, Reliability and Quality Assurance,* 1994, pp. 68–72.
468 Ramamoorthy, C.V., "Evolution and Evaluation of Software Quality Models," *Proceedings of the 14th International Conference on Tools with Artificial Intelligence,* 2002, pp. 543–546.
469 Raman, S., "CMM: A Road Map to Software Quality," *Proceedings of the 51st Annual Quality Congress,* 1997, pp. 898–906.
470 Ramanna, S., "Approximation Methods in a Software Quality Measurement Framework," *Proceedings of the IEEE Canadian Conference on Electrical and Computer Engineering,* 2002, pp. 566–571.
471 Ramanna, S., Peters, J.F., Ahn, T., "Software Quality Knowledge Discovery: A Rough Set Approach," *Proceedings of the 26th Annual International Computer Software and Applications Conference,* 2002, pp. 1140–1145.
472 Redig, G., Swanson, M., "Control Data Corporation's Government Systems Group Standard Software Quality Program," *Proceedings of the IEEE National Aerospace and Electronics Conference*, 1990, pp. 670–674.
473 Redig, G., Swanson, M., "Total Quality Management for Software Development," *Proceedings of the 6th Annual IEEE Symposium on Computer-Based Medical Systems,* 1993, pp. 301–306.
474 Redmill, F.J., "Considering Quality in the Management of Software-based Development Projects," *Information and Software Technology,* Vol. 32, No. 1, 1990, pp. 18–25.
475 Rigby, P.J., Stoddart, A.G., Norris, M.T., "Assuring Quality in Software – Practical Experiences in Attaining ISO 9001," *British Telecommunications Engineering,* Vol. 8, No. 4, 1990, pp. 244–249.
476 Rodford, J., "New Levels of Software Quality by Automatic Software Inspection," *Electronic Engineering (London),* Vol. 72, No. 882, 2000, pp. 3–9.
477 Rosenberg, L.H., Sheppard, S.B., "Metrics in Software Process Assessment, Quality Assurance and Risk Assessment," *Proceedings of the 2nd International Software Metrics Symposium,* 1994, pp. 10–16.
478 Rozum, J.A., "Roadmap to Improving Software Productivity and Quality," *Proceedings of the ISA TECH/EXPO Technology Update Onference*, 1997, pp. 185–194.
479 Russell, B., Chatterjee, S., "Relationship Quality: The Undervalued Dimension of Software Quality," *Communications of the ACM,* Vol. 46, No. 8, 2003, pp. 85–89.
480 Sakasai, Y., Hotta, K., "Construction of Software Quality Assurance System," *NTT R&D,* Vol. 45, No. 3, 1996, pp. 237–246.
481 Salameh, W.A., "Comparative Study on Software Quality Assurance using Complexity Metrics," *Advances in Modelling and Analysis A,* Vol. 22, No. 1, 1994, pp. 1–6.
482 Scerbo, M.W., "Usability Engineering Approach to Software Quality," *Proceedings of the 45th Annual Quality Congress,* 1991, pp. 726–733.
483 Schneidewind, N.F., "Body of Knowledge for Software Quality Measurement," *Computer,* Vol. 35, No. 2, 2002, pp. 77–83.
484 Schneidewind, N.F., "Knowledge Requirements for Software Quality Measurement," *Empirical Software Engineering,* Vol. 6, No. 3, 2001, pp. 201–205.

485 Schneidewind, N.F., "Report on the IEEE Standard for a Software Quality Metrics Methodology," *Proceedings of the Conference on Software Maintenance,* 1993, pp. 104–106.
486 Schneidewind, N.F., "Software Metrics Model for Integrating Quality Control and Prediction," *Proceedings of the 8th International Symposium on Software Reliability Engineering,* 1997, pp. 402–415.
487 Schneidewind, N.F., "Software Metrics Model for Quality Control," *Proceedings of the 4th International Software Metrics Symposium,* 1997, pp. 127–136.
488 Schneidewind, N.F., "Software Quality Maintenance Model," *Proceedings of the Conference on Software Maintenance,* 1999, pp. 277–286.
489 Schoonmaker, S.J., "Engineering Software Quality Management," *Proceedings of the Energy-Sources Technology Conference and Exhibition,* 1992, pp. 55–63.
490 Scicchitano, P., "Great ISO 9000 Debate: Quality in Software," *Compliance Engineering,* Vol. 12, No. 7, 1995, pp. 4–5.
491 Sedigh-Ali, S., Ghafoor, A., Paul, R.A., "Metrics-guided Quality Management for Component-based Software Systems," *Proceedings of the 25th Annual International Computer Software and Applications Conference,* 2001, pp. 303–308.
492 Shepperd, M., "Early Life-cycle Metrics and Software Quality Models," *Information and Software Technology,* Vol. 32, No. 4, 1990, pp. 311–316.
493 Sherif, Y.S., Kelly, J.C., "Improving Software Quality through Formal Inspections," *Microelectronics and Reliability,* Vol. 32, No. 3, 1992, pp. 423–431.
494 Shi, X., Wang, H., Zhong, Q., "Design and Implementation of a Software Quality Evaluation System," *Jisuanji Gongcheng/Computer Engineering,* Vol. 24, No. 2, 1998, pp. 8–11.
495 Simmons, R.A., "Software Quality Assurance (SQA) Early in the Acquisition Process," *Proceedings of the IEEE National Aerospace and Electronics Conference,* 1990, pp. 664–669.
496 Slaughter, S.A., Harter, D.F., and Krishna, M.S., "Evaluating the Cost of Software Quality," *IEEE Engineering Management Review,* Vol. 26, No. 4, 1998, pp. 32–37.
497 Sova, D.W., Smidts, C., "Increasing Testing Productivity and Software Quality: A Comparison of Software Testing Methodologies within NASA," *Empirical Software Engineering,* Vol. 1, No. 2, 1996, pp. 165–188.
498 Spinelli, A., Pina, D., Salvaneschi, P., "Quality Measurement of Software Products. An Experience about a Large Automation System," *Proceedings of the 2nd Symposium on Software Quality Techniques and Acquisition Criteria,* 1995, pp. 192–194.
499 Squires, D., Preece, J., "Predicting Quality in Educational Software: Evaluating for Learning, Usability and the Synergy between them," *Interacting with Computers,* Vol. 11, No. 5, 1999, pp. 467–483.
500 Staknis, M.E., "Software Quality Assurance through Prototyping and Automated Testing," *Information and Software Technology,* Vol. 32, No. 1, 1990, pp. 26–33.
501 Stieber, H.A., "Statistical Quality Control: How to Detect Unreliable Software Components," *Proceedings of the 8th International Symposium on Software Reliability Engineering,* 1997, pp. 8–12.
502 Stockman, S., "Total Quality Management in Software Development," *Proceedings of the IEEE Global Telecommunications Conference,* 1993, pp. 498–504.
503 Stockman, S.G., Todd, A.R., Robinson, G.A., "Framework for Software Quality Measurement," *IEEE Journal on Selected Areas in Communications,* Vol. 8, No. 2, 1990, pp. 224–233.

504 Suri, D., "Software Quality Assurance for Software Engineers," *Proceedings of the ASEE Annual Conference and Exposition,* 2004, pp. 12727–12735.
505 Suryn, W., Abran, A., April, A., "ISO/IEC SQuaRE. The Second Generation of Standards for Software Product Quality," *Proceedings of the Seventh IASTED International Conference on Software Engineering and Applications,* 2003, pp. 807–814.
506 Takahashi, R., "Software Quality Classification Model based on McCabe's Complexity Measure," *Journal of Systems and Software,* Vol. 38, No. 1, 1997, pp. 61–69.
507 Takahashi, R., Muraoka, Y., Nakamura, Y., "Building Software Quality Classification Trees: Approach, Experimentation, Evaluation," *Proceedings of the 8th International Symposium on Software Reliability Engineering,* 1997, pp. 222–233.
508 Talbert, N.B., "Representative Sampling within Software Quality Assurance," *Proceedings of the Conference on Software Maintenance,* 1993, pp. 174–179.
509 Tanaka, T., Aizawa, M., Ogasawara, H., "Software Quality Analysis & Measurement Service Activity in the Company," *Proceedings of the International Conference on Software Engineering,* 1998, pp. 426–429.
510 Tervonen, I., "Necessary Skills for a Software Quality Engineer," *Proceedings of the 2nd International Conference on Software Quality Management*, 1994, pp. 573–577.
511 Tervonen, I., and Kerola, P., "Towards Deeper Co-understanding of Software Quality," *Information and Software Technology,* Vol. 39, No. 14–15, 1998, pp. 995–1003.
512 Thompson, J.B., Edwards, H.M., "STePS. A method that will Improve Software Quality," *Proceedings of the 2nd International Conference on Software Quality Management*, 1994, pp. 131.
513 Tian, J., "Early Measurement and Improvement of Software Quality," *Proceedings of the IEEE 22nd Annual International Computer Software & Applications Conference,* 1998, pp. 196–201.
514 Tian, J., Software Quality Engineering: Testing, Quality Assurance, and Quantifiable Improvement, Wiley, New York, 2005.
515 Tiwari, A., Tandon, A., "Shaping Software Quality – The Quantitative Way," *Proceedings of the 1st International Conference on Software Testing, Reliability and Quality Assurance,* 1994, pp. 84–94.
516 Todd, A., "Measurement of Software Quality Improvement," *Proceedings of the Colloquium on Software Metrics,* 1990, pp. 8.1–8.5.
517 Toyama, M., Sugawara, M., Nakamura, K., "High-quality Software Development System – AYUMI," *IEEE Journal on Selected Areas in Communications,* Vol. 8, No. 2, 1990, pp. 201–209.
518 Trammell, C.J., Poore, J.H., "Group Process for Defining Local Software Quality. Field Applications and Validation Experiments," *Software – Practice and Experience,* Vol. 22, No. 8, 1992, pp. 603–636.
519 Ueyama, Y., Ludwig, C., "Joint Customer Development Process and its Impact on Software Quality," *IEEE Journal on Selected Areas in Communications,* Vol. 12, No. 2, 1994, pp. 265–270.
520 Visaggio, G., "Structural Information as a Quality Metric in Software Systems Organization," *Proceedings of the International Conference on Software Maintenance,* 1997, pp. 92–99.

521 Voas, J., "Assuring Software Quality Assurance," *IEEE Software,* Vol. 20, No. 3, 2003, pp. 48–49.
522 Voas, J., "The COTS Software Quality Challenge," *Proceedings of the 56th Annual Quality Congress*, 2002, pp. 93–96.
523 Voas, J., Agresti, W.W., "Software Quality from a Behavioral Perspective," *IT Professional,* Vol. 6, No. 4, 2004, pp. 46–50.
524 Vollman, T., "Standards Support for Software Tool Quality Assessment," *Proceedings of the 3rd Symposium on Assessment of Quality Software Development Tools,* 1994, pp. 29–39.
525 Von Hellens, L.A., "Information Systems Quality Versus Software Quality – A Discussion From a Managerial, an Organizational and an Engineering Viewpoint," *Information and Software Technology,* Vol. 39, No. 12, 1997, pp. 801–808.
526 Wackerbarth, G., "Quality Assurance for Software," *Forensic Engineering,* Vol. 2, No. 1–2, 1990, pp. 97–111.
527 Waite, D.A., "Software Quality Management from the Outside In," *Proceedings of the ASQC Annual Quality Congress,* 1994, pp. 778–782.
528 Walker, A.J., "Improving the Quality of ISO 9001 Audits in the Field of Software," *Information and Software Technology,* Vol. 40, No. 14, 1998, pp. 865–869.
529 Walker, A.J., "Quality Management Applied to the Development of a National Checklist for ISO 9001 Audits for Software," *Proceedings of the 3rd IEEE International Software Engineering Standards Symposium and Forum,* 1997, pp. 6–14.
530 Wallmueller, E., "Software Quality Management," *Microprocessing and Microprogramming,* Vol. 32, No. 1–5, 1991, pp. 609–616.
531 Walsh, J., "Software Quality Management: A Subjective View," *Engineering Management Journal,* Vol. 4, No. 3, 1994, pp. 105–111.
532 Walton, T., "Quality Planning for Software Development," *Proceedings of the 25th Annual International Computer Software and Applications Conference*, 2001, pp. 104–109.
533 Wegner, E., "Quality of software packages: The Forthcoming International Standard," *Computer Standards & Interfaces,* Vol. 17, No. 1, 1995, pp. 115–120.
534 Wells, C.H., Brand, R., Markosian, L., "Customized Tools for Software Quality Assurance and Reengineering," *Proceedings of the 2nd Working Conference on Reverse Engineering,* 1995, pp. 71–77.
535 Welzel, D., Hausen, H., "Five Step Method for Metric-based Software Evaluation – Effective Software Metrication with Respect to Quality Standards," *Microprocessing and Microprogramming,* Vol. 39, No. 2–5, 1993, pp. 273–276.
536 Werth, L.H., "Quality Assessment on a Software Engineering Project," *IEEE Transactions on Education,* Vol. 36, No. 1, 1993, pp. 181–183.
537 Wesenberg, D.P., Vansaun, K., "A System Approach for Software Quality Assurance," *Proceedings of the IEEE National Aerospace and Electronics Conference*, 1991, pp. 771–776.
538 Wesselius, J., Ververs, F., "Some Elementary Questions on Software Quality Control," *Software Engineering Journal,* Vol. 5, No. 6, 1990, pp. 319–330.
539 Weyuker, E.J., "Evaluation Techniques for Improving the Quality of Very Large Software Systems in a Cost-effective Way," *Journal of Systems and Software,* Vol. 47, No. 2, 1999, pp. 97–103.
540 Wheeler, S., Duggins, S., "Improving Software Quality," *Proceedings of the 36th Annual Southeast Conference,* 1998, pp. 300–309.

541 Whittaker, J.A., Voas, J.M., "50 Years of Software: Key Principles for Quality," *IT Professional,* Vol. 4, No. 6, 2002, pp. 28–35.
542 Wilson, D., "Software Quality Assurance in Australia," *Proceedings of the International Conference on Software Quality Management,* 1993, pp. 911–914.
543 Wong, B., "Measurements used in Software Quality Evaluation," *Proceedings of the International Conference on Software Engineering Research and Practise,* 2003, pp. 971–977.
544 Woodman, M., "Making Software Quality Assurance a Hidden Agenda?" *Proceedings of the International Conference on Software Quality Management,* 1993, pp. 301–305.
545 Wu, C., Lin, J., Yu, L., "Research of Military Software Quality Characteristic and Design Attribute," *Jisuanji Gongcheng/Computer Engineering,* Vol. 31, No. 12, 2005, pp. 100–102.
546 Xenos, M., Christodoulakis, D., "Applicable Methodology to Automate Software Quality Measurements," *Proceedings of the International Conference on Software Testing, Reliability and Quality Assurance,* 1994, pp. 121–125.
547 Xiao-Ying, S., Lan, Y., "A Software Quality Evaluation System: JT-SQE," *Wuhan University Journal of Natural Sciences,* Vol. 6, No. 1–2, 2001, pp. 511–515.
548 Xu, Z., Khoshgoftaar, T.M., "Software Quality Prediction for High-assurance Network Telecommunications Systems," *Computer Journal,* Vol. 44, No. 6, 2001, pp. 558–568.
549 Yamada, S., "Software Quality/reliability Measurement and Assessment: Software Reliability Growth Models and Data Analysis," *Journal of Information Processing,* Vol. 14, No. 3, 1991, pp. 254–266.
550 Yan, H., Hu, J., Zhang, L., "Process Management for Software Quality Assurance Based on Document Status," *Beijing Hangkong Hangtian Daxue Xuebao/Journal of Beijing University of Aeronautics and Astronautics,* Vol. 27, No. 4, 2001, pp. 474–477.
551 Yang, Y., "Synthetic Evaluation Method for Software Quality," *Xiaoxing Weixing Jisuanji Xitong/Mini-Micro Systems,* Vol. 21, No. 3, 2000, pp. 313–315.
552 Yang, Y.H., "Software Quality Management and ISO 9000 Implementation," *Industrial Management and Data Systems,* Vol. 101, No. 7, 2001, pp. 329–338.
553 Yau, S.S., Wang, Y., Huang, J.G., "An Integrated Expert System Framework for Software Quality Assurance," *Proceedings of the 14th Annual International Computer Software and Applications Conference,* 1990, pp. 161–166.
554 Yokoyama, Y., Kodaira, M., "Software Cost and Quality Analysis by Statistical Approaches," *Proceedings of the International Conference on Software Engineering,* 1998, pp. 465–467.
555 Zargari, A., Cantrell, P.A., Grise, W., "Application of Statistical Process Control (SPC) Software in Total Quality Improvement (TQI)," *Proceedings of the 23rd IEEE Electrical Electronics Insulation Conference and Electrical Manufacturing & Coil Winding,* 1997, pp. 829–834.
556 Zeng, X., Tsai, J.J.P., Weigert, T.J., "Improving Software Quality through a Novel Testing Strategy," *Proceedings of the 19th Annual International Computer Software and Applications Conference,* 1995, pp. 224–229.
557 Zhang, S., Liu, X., Deng, Y., "Software Quality Metrics Methodology and its Application," *Beijing Hangkong Hangtian Daxue Xuebao/Journal of Beijing University of Aeronautics and Astronautics,* Vol. 23, No. 1, 1997, pp. 61–67.

558 Zhu, H., Zhang, Y., Huo, Q., "Application of Hazard Analysis to Software Quality Modelling," *Proceedings of the 26th Annual International Computer Software and Applications Conference,* 2002, pp. 139–144.
559 Zultner, R.E., "Software Quality Function Deployment: The First Five Years – Lessons Learned," *Proceedings of the ASQC 48th Annual Quality Congress,* 1994, pp. 783–793.
560 Zweben, S.H., "Evaluating the Quality of Software Quality Indicators," *Proceedings of the Twenty-Second Symposium on the Interface, Computing Science and Statistics,* 1992, pp. 266–269.

A.2.6 Robot Reliability

561 Ayrulu, B., Barshan, B., "Raliability Measure Assignment to Sonar for Robust Target Differentiation", *Pattern Recognition,* Vol. 35, No. 6, 2002, pp. 1403–1419.
562 Becker, C., Salas, J., Tokusei, K., Latombe, J., "Reliable Navigation using Landmarks", *Proceedings of the IEEE International Conference on Robotics and Automation,* Vol. 1, 1995, pp. 401–406.
563 Brooks, R.R., Iyengar, S.S., "Robot Algorithm Evaluation by Simulating Sensor Faults", *Proceedings of SPIE Conference*, Vol. 2484, 1995, pp. 394–401.
564 Carlson, J., Murphy, R.R., Nelson, A., "Follow-Up Analysis of Mobile Robot Failures", *Proceedings of the IEEE International Conference on Robotics and Automation,* Vol. 2004, No. 5, 2004, pp. 4987–4994.
565 Carreras, C., Walker, I.D., "Interval Methods for Fault-Tree Analysis in Robotics", *IEEE Transactions on Reliability,* Vol. 50, No. 1, 2001, pp. 3–11.
566 Carreras, C., Walker, I.D., "Interval Methods for Improved Robot Reliability Estimation", *Proceedings of the Annual Reliability and Maintainability Symposium,* 2000, pp. 22–27.
567 Carreras, C., Walker, I.D., "On Interval Methods Applied to Robot Reliability Quantification", *Reliability Engineering and System Safety,* Vol. 70, No. 3, 2000, pp. 291–303.
568 Chen, I., "Effect of Parallel Planning on System Reliability of Real-Time Expert Systems", *IEEE Transactions on Reliability,* Vol. 46, No. 1, 1997, pp. 81–87.
569 Chen, W., Hou, L., Cai, H., "Research on Reliability Distribution Method for Portable Robot System", *Jiqiren/Robot,* Vol. 24, No. 1, 2002, pp. 35–38.
570 Chen, W., Hou, L., Cai, H., "Two-Level Reliability Optimizing Distribution and its Algorithms for Series Industry Robot Systems", *Gaojishu Tongxin/High Technology Letters,* Vol. 10, No. 2, 2000, pp. 78–81.
571 Chen, W., Hou, L., Cai, Hegao, "Reliability Optimizing Distribution for Portable Arc-Welding Robot System", *Harbin Gongye Daxue Xuebao/Journal of Harbin Institute of Technology,* Vol. 32, No. 1, 2000, pp. 12–14.
572 Chinellato, E., Morales, A., Fisher, R.B., Del Pobil, A.P., "Visual Quality Measures for Characterizing Planar Robot Grasps", *IEEE Transactions on Systems, Man and Cybernetics Part C: Applications and Reviews,* Vol. 35, No. 1, 2005, pp. 30–41.
573 Dhillon, B.S., Aleem, M.A., "Report on Robot Reliability and Safety in Canada: A Survey of Robot Users", *Journal of Quality in Maintenance Engineering,* Vol. 6, No. 1, 2000, pp. 61–74.

574 Dhillon, B.S., Anude, O.C., "Robot Safety and Reliability: A Review", *Microelectronics and Reliability,* Vol. 33, No. 3, 1993, pp. 413–429.
575 Dhillon, B.S., Fashandi, A.R.M., "Robotic Systems Probabilistic Analysis", *Microelectronics and Reliability,* Vol. 37, No. 2, 1997, pp. 211–224.
576 Dhillon, B.S., Fashandi, A.R.M., "Safety and Reliability Assessment Techniques in Robotics", *Robotica,* Vol. 15, No.6, 1997, pp. 701–708.
577 Dhillon, B.S., Fashandi, A.R.M., "Stochastic Analysis of a Robot Machine with Duplicate Safety Units", *Journal of Quality in Maintenance Engineering,* Vol. 5, No. 2, 1999, pp. 114–127.
578 Dhillon, B.S., Fashandi, A.R.M., Liu, K.L., "Robot Systems Reliability and Safety: A Review", *Journal of Quality in Maintenance Engineering,* Vol. 8, No. 3, 2002, pp. 170–212.
579 Dhillon, B.S., Li, Z., "Stochastic Analysis of a Maintainable Robot-Safety System with Common-Cause Failures", *Journal of Quality in Maintenance Engineering,* Vol. 10, No. 2, 2004, pp. 136–147.
580 Dhillon, B.S., Yang, N., "Formulas for Analyzing a Redundant Robot Configuration with a Built-in Safety System", *Microelectronics and Reliability,* Vol. 37, No. 4, 1997, pp. 557–563.
581 Dhillon, B.S., Yang, N., "Reliability Analysis of a Repairable Robot System", *Journal of Quality in Maintenance Engineering,* Vol. 2, No. 2, 1996, pp. 30–37.
582 Dixon, W.E., Walker, I.D., Dawson, D.M., Hartranft, J.P., "Fault Detection for Robot Manipulators with Parametric Uncertainty: A Prediction Error Based Approach", *Proceedings of the IEEE International Conference on Robotics and Automation,* Vol. 4, 2000, pp. 3628–3634.
583 Ehrenweber, R., "Testing Quality Characteristics with a Robot", *Kunststoffe Plast Europe,* Vol. 86, No. 4, 1996, pp. 12–13.
584 Friedrich, W.E., "Robotic Handling: Sensors Increase Reliability", *Industrial Robot,* Vol. 22, No. 4, 1995, pp. 23–26.
585 Gu, Y., Zhou, H., Ma, H., "Group-Control on Multiped Robot and Method of Reliability", *Jiqiren/Robot,* Vol. 24, No. 2, 2002, pp. 140–144.
586 Guiochet, J., Tondu, B., Baron, C., "Integration of UML in Human Factors Analysis for Safety of a Medical Robot for Tele-Echography", *Proceedings of the IEEE International Conference on Intelligent Robots and Systems*, Vol. 4, 2003, pp. 3212–3217.
587 Hamilton, D.L., Walker, I.D., Bennett, J.K., "Fault Tolerance versus Performance Metrics for Robot Systems", *Reliability Engineering & System Safety,* Vol. 53, No. 3, 1996, pp. 309–318.
588 Harpel, B.M., Dugan, J.B., Walker, I.D., Cavallaro, J.R., "Analysis of Robots for Hazardous Environments", *Proceedings of the Annual Reliability and Maintainability Symposium,* 1997, pp. 111–116.
589 Hosaka, S., Shimizu, Y., Hayashi, T., "Development of a Functional Fail-Safe Control for Advanced Robots", *Advanced Robotics,* Vol. 8, No. 5, 1994, pp. 477–495.
590 Kanatani, K., Ohta, N., "Optimal Robot Self-Localization and Reliability Evaluation", *Lecture Notes in Computer Science,* Vol. 1407, 1998, pp. 796–800.
591 Khodabandehloo, K., "Analyses of Robot Systems using Fault and Event Trees: Case Studies", *Reliability Engineering & System Safety,* Vol. 53, No. 3, 1996, pp. 247–264.

592 Kobayashi, F., Arai, F., Fukuda, T., "Sensor Selection by Reliability Based on Possibility Measure", *Proceedings of the IEEE International Conference on Robotics and Automation,* Vol. 4, 1999, pp. 2614–2619.
593 Koker, R., "Reliability-Based Approach to the Inverse Kinematics Solution of Robots using Elman's Networks", *Engineering Applications of Artificial Intelligence,* Vol. 18, No. 6, 2005, pp. 685–693.
594 Lauridsen, K., "Reliability of Remote Manipulator Systems for use in Radiation Environments", *Proceedings of the Computing and Control Division Colloquium on Safety and Reliability of Complex Robotic Systems*, 1994, pp. 1.1–1.5.
595 Lee, S., Choi, D.S., Kim, M., Lee, C.W., Song, J.B., "Human and Robot Integrated Teleoperation", *Proceedings of the IEEE International Conference on Systems, Man, and Cybernetics*, Vol. 2, 1998, pp. 1213–1218.
596 Lendvay, M., "Accelerating Reliability Test of Electromechanical Contacts to Robot Controlling", *Proceedings of the International Conference on Intelligent Engineering Systems,* 1997, pp. 421–425.
597 Leuschen, M.L., Walker, I.D., Cavallaro, J.R., "Evaluating the Reliability of Prototype Degradable Systems", *Reliability Engineering and System Safety,* Vol. 72, No. 1, 2001, pp. 9–20.
598 Leuschen, M.L., Walker, I.D., Cavallaro, J.R., "Robot Reliability through Fuzzy Markov Models", *Proceedings of the Annual Reliability and Maintainability Symposium,* 1998, pp. 209–214.
599 Lewis, C.L., Maciejewski, A.A., "Example of Failure Tolerant Operation of a Kinematically Redundant Manipulator", *Proceedings of the IEEE International Conference on Robotics and Automation,* 1994, pp. 1380–1387.
600 Lin, C., Wang, M.J., "Hybrid Fault Tree Analysis using Fuzzy Sets", *Reliability Engineering & System Safety,* Vol. 58, No. 3, 1997, pp. 205–213.
601 Liu, T.S., Wang, J.D., "Reliability Approach to Evaluating Robot Accuracy Performance", *Mechanism & Machine Theory,* Vol. 29, No. 1, 1994, pp. 83–94.
602 Lueth, T.C., Nassal, U.M., Rembold, U., "Reliability and Integrated Capabilities of Locomotion and Manipulation for Autonomous Robot Assembly", *Robotics and Autonomous Systems,* Vol. 14, No. 2–3, 1995, pp. 185–198.
603 McInroy, J.E., Saridis, G., "Entropy Searches for Robotic Reliability Assessment", *Proceedings of the IEEE International Conference on Robotics and Automation,* Vol. 1, 1993, pp. 935–940.
604 Michaelson, D.G., Jiang, J., "Modelling of Redundancy in Multiple Mobile Robots", *Proceedings of the American Control Conference*, Vol. 2, 2000, pp. 1083–1087.
605 Moon, I., Joung, S. Kum, Y., "Safe and Reliable Intelligent Wheelchair Robot with Human-Robot Interaction", *Proceedings of the IEEE International Conference on Robotics and Automation,* Vol. 4, 2002, pp. 3595–3600.
606 Morales, A., Chinelalto, E., Sanz, P.J., Del Pobil, A.P., Fagg, A.H., "Learning to Predict Grasp Reliability for a Multifinger Robot Hand by using Visual Features", *Proceedings of the Eighth IASTED International Conference on Atificial Intelligence and Soft Computing,* 2004, pp. 249–254.
607 Morales, A., Chinellato, E., Fagg, A.H., Del Pobil, A.P., "An Active Learning Approach for Assessing Robot Grasp Reliability", *Proceedings of the IEEE/RSJ International Conference on Intelligent Robots and Systems (IROS)*, Vol. 1, 2004, pp. 485–490.

608 Narita, S., Ohkami, Y., "Development of Distributed Controller Software for Improving Robot Performance and Reliability", *Proceedings of the IEEE/RSJ International Conference on Intelligent Robots and Systems (IROS)*, Vol. 3, 2004, pp. 2384–2389.
609 Ntuen, C.A., Park, E.H., "Formal Method to Characterize Robot Reliability", *Proceedings of the Annual Reliability and Maintainability Symposium*, 1993, pp. 395–397.
610 Okina, S., Kawabata, K., Fujii, T., Kunii, Y., Asama, H., Endo, I., "Study of a Self-Diagnosis System for an Autonomous Mobile Robot", *Advanced Robotics*, Vol. 14, No. 5, 2000, pp. 339–341.
611 Otsuka, K., "Inspection Robot System of Electronic Unit Manufacturing", *Robot*, No. 111, 1996, pp. 22–27.
612 Pirjanian, P., "Reliable Reaction", *Proceedings of the IEEE/SICE/RSJ International Conference on Multisensor Fusion and Integration for Intelligent Systems*, 1996, pp. 158–165.
613 Pomerleau, D.A., "Reliability Estimation for Neural Network Based Autonomous Driving", *Robotics and Autonomous Systems*, Vol. 12, No. 3–4, 1994, pp. 113–119.
614 Ramachandran, S., Nagarajan, T., Sivaprasad, N., "Reliability Studies on Assembly Robots using the Finite Element Method", *Advanced Robotics*, Vol. 7, No. 4, 1993, pp. 385–393.
615 Renaud, P., Cervera, E., Martinet, P., "Towards a Reliable Vision-Based Mobile Robot Formation Control", *Proceedings of the IEEE/RSJ International Conference on Intelligent Robots and Systems*, Vol. 4, 2004, pp. 3176–3181.
616 Roston, G.P., Dowling, K., "Drivetrain Design, Incorporating Redundancy, for an Autonomous Walking Robot", *Proceedings of the ASCE Specialty Conference on Robotics for Challenging Environments Conference*, 1994, pp. 184–192.
617 Savsar, M., "Reliability Analysis of a Flexible Manufacturing Cell", *Reliability Engineering & System Safety*, Vol. 67, No. 2, 2000, pp. 147–152.
618 Shi, T., Zhang, Z., Wang, X., Zhu, X., "Approach to Fault on-Line Detection and Diagnosis Based on Neural Networks for Robot in FMS", *Chinese Journal of Mechanical Engineering (English Edition)*, Vol. 11, No. 2, 1998, pp. 115–121.
619 Smrcek, J., Neupauer, R., "Testing of Intelligent Robots, Development and Experience", *Proceedings of the IEEE International Conference on Intelligent Engineering Systems*, 1997, pp. 119–121.
620 Suita, K., Yamada, Y., Tsuchida, N., Imai, K., Ikeda, H., Sugimoto, N., "Failure-to-Safety 'Kyozon' System with Simple Contact Detection and Stop Capabilities for Safe Human-Autonomous Robot Coexistence", *Proceedings of the IEEE International Conference on Robotics and Automation*, Vol. 3, 1995, pp. 3089–3096.
621 Tadokoro, S., Takebe, T., Ishikawa, Y., Takamori, T., "Stochastic Prediction of Human Motion and Control of Robots in the Service of Human", *Proceedings of the International Conference on Systems, Man and Cybernetics*, Vol. 1, 1993, pp. 503–508.
622 Taogeng, Z., Dingguo, S. Qun, H., "Application of Bayes Data Fusion in Robot Reliability Assessment", *Proceedings of the International Workshop on Bio-Robotics and Teleoperation*, 2001, pp. 262–265.
623 Tinos, R., Navarro-Serment, L.E., Paredis, C.J.J., "Fault Tolerant Localization for Teams of Distributed Robots", *Proceedings of the International Conference on Intelligent Robot and Systems*, 2001, pp. 1061–1066.

624 Toye, G., Leifer, L.J., "Helenic Fault Tolerance for Robots", *Computers & Electrical Engineering,* Vol. 20, No. 6, 1994, pp. 479–497.
625 Vanderperre, E.J., Makhanov, S.S., "Risk Analysis of a Robot-Safety Device System", *International Journal of Reliability, Quality and Safety Engineering,* Vol. 9, No. 1, 2002, pp. 79–87.
626 Walker, I.D., Cavallaro, J.R., "Use of Fault Trees for the Design of Robots for Hazardous Environments", *Proceedings of the Annual Reliability and Maintainability Symposium,* 1996, pp. 229–235.
627 Wikman, T.S., Branicky, M.S., Newman, W.S., "Reflex Control for Robot System Preservation, Reliability and Autonomy", *Computers & Electrical Engineering,* Vol. 20, No. 5, 1994, pp. 391–407.
628 Willetts, N., "Jaguar. A Robot User's Experience", *Industrial Robot,* Vol. 20, No. 1, 1993, pp. 9–12.
629 Wilson, M.S., "Reliability and Flexibility – a Mutually Exclusive Problem for Robotic Assembly?", *IEEE Transactions on Robotics and Automation,* Vol. 12, No. 2, 1996, pp. 343–347.
630 Xu, F., Deng, Z., "The Reliability Apportionment for X-Ray Inspection Real Time Imaging Pipeline Robot Based on Fuzzy Synthesis", *Proceedings of the Fifth World Congress on Intelligent Control and Automation,* 2004, pp. 4763–4767.
631 Yamada, A., Takata, S., "Reliability Improvement of Industrial Robots by Optimizing Operation Plans Based on Deterioration Evaluation", *CIRP Annals – Manufacturing Technology,* Vol. 51, No. 1, 2002, pp. 319–322.
632 Yamada, Y., Suita, K., Imai, K., Ikeda, H., Sugimoto, N., "Failure-to-Safety Robot System for Human-Robot Coexistence", *Robotics and Autonomous Systems,* Vol. 18, No. 1–2, 1996, pp. 283–291.
633 Yang, M., Wang, H., He, K. Zhang, B., "Environmental Modeling and Obstacle Avoidance of Mobile Robots Based on Laser Radar", *Qinghua Daxue Xuebao/Journal of Tsinghua University,* Vol. 40, No. 7, 2000, pp. 112–116.
634 Yu, S., McCluskey, E.J., "On-Line Testing and Recovery in TMR Systems for Real-Time Applications", *Proceedings of the International Test Conference,* 2001, pp. 240–249.
635 Zhang, J., Lu, T., "Fuzzy Fault Tree Analysis of a Cable Painting Robot", *Shanghai Jiaotong Daxue Xuebao/Journal of Shanghai Jiaotong University,* Vol. 37, No.5, 2003, pp. 62–65.
636 Zhou, G., "Position and Orientation Calibration and Moving Reliability of the Robot used in Pyrometallurgy Process", *Zhongnan Gongye Daxue Xuebao/Journal of Central South University of Technology,* Vol. 31, No. 6, 2000, pp. 556–560.

A.2.7 Power System Reliability

637 Abdallah, E.N.,"A Novel Expert System to Calculate Reliability and Bus Failure Frequency of Composite Power Systems Based on the Tearing Process," *AEJ – Alexandria Engineering Journal,* Vol. 43, No. 6, 2004, pp. 791–800.
638 Abdelaziz, A.R.,"Fast Algorithm for Network Decomposition and its Application to Reliability of Large Power Systems," *AEJ – Alexandria Engineering Journal,* Vol. 34, No. 5, 1995, pp. 139–144.

639 Abdelaziz, A.R.,"Fuzzy-Based Power System Reliability Evaluation," *Electric Machines and Power Systems,* Vol. 27, No. 3, 1999, pp. 271–278.
640 Abdelaziz, A.R.,"Reliability Evaluation in Operational Planning of Power Systems," *Electric Machines and Power Systems,* Vol. 25, No. 4, 1997, pp. 419–428.
641 Ahsan, Q., Bokshi, M.A., "Appropriate Method of Evaluating Reliability of Small Power Systems. A Case Study," *Journal of the Institution of Engineers (India), Part EL: Electrical Engineering Division,* Vol. 72, No. 1, 1991, pp. 7–10.
642 Alavi-Serashki, M.M., Singh, C., "Generalized Cumulant Method for Generation Capacity Reliability Evaluation," *Electric Power Systems Research,* Vol. 23, No. 1, 1992, pp. 1–4.
643 Albaugh, J.R., Kornblit, M.J., "Updating a Steel Mill Power System to Improve Reliability and Decrease Energy Costs," *Proceedings of the 28th Annual Meeting of the IEEE Industry Applications Conference,* Vol. 3, 1993, pp. 2415–2419.
644 Allan, R., Billinton, R. "Probabilistic Assessment of Power Systems," *Proceedings of the IEEE,* Vol. 88, No. 2, 2000, pp. 140–162.
645 Allan, R.N., Bhuiyan, M.R., "Application of Sequential Simulation to the Reliability Assessment of Bulk Power Systems," *Proceedings of the 29th Universities Power Engineering Conference*, Vol. 2, 1994, pp. 763–766.
646 Allan, R.N., Billinton, R., Sjarief, I., Goel, L., So, K.S., "A Reliability Test System for Educational Purposes – Basic Distribution System Data and Results," *IEEE Transactions on Power Systems,* Vol. 6, No. 2, 1991, pp. 813–820.
647 Allan, R.N., Da Silva, M.G. "Evaluation of Reliability Indices and Outage Costs in Distribution Systems," *IEEE Transactions on Power Systems,* Vol. 10, No. 1, 1995, pp. 413–419.
648 Allen, C.W., Colocho, W., Erickson, R., Stanek, M., "PEP-II Hardware Reliability," *Proceedings of the IEEE Nuclear Science Symposium*, Vol. 3, 2004, pp. 1788–1792.
649 Allen, E.H., Ilic, M.D. "Reserve Markets for Power Systems Reliability," *IEEE Transactions on Power Systems,* Vol. 15, No. 1, 2000, pp. 228–233.
650 Amjady, N., Ehsan, M. "Evaluation of Power Systems Reliability by an Artificial Neural Network," *IEEE Transactions on Power Systems,* Vol. 14, No. 1, 1999, pp. 287–292.
651 Amjady, N., Farrokhzad, D., Modarres, M. "Optimal Reliable Operation of Hydrothermal Power Systems with Random Unit Outages," *IEEE Transactions on Power Systems,* Vol. 18, No. 1, 2003, pp. 279–287.
652 Anishchenko, V.A.,"Monitoring of Measurement Reliability in Power Systems on the Basis of Statistical Decision Theory," *Izvestiya Vysshikh Uchebnykh Zavedenij i Energeticheskikh Ob''edinenij Sng. Energetika*, No. 6, 2003, pp. 5–15.
653 Antonov, G.N., Voropaj, N.I., Krivorutskij, L.D., et al, "Complex Investigations for Survivability of Power Systems," *Izvestiya Akademii Nauk.Energetika,* No. 6, 1992, pp. 31–41.
654 Asgarpoor, S. and Singh, C., "Analytical Technique for Bulk Power System Reliability Evaluation," *Electric Power Systems Research,* Vol. 20, No. 1, 1990, pp. 63–71.
655 Asgarpoor, S., Singh, C., "New Index for Generation Capacity Reliability Studies: Expected Cost Penalty," *Electric Power Systems Research,* Vol. 23, No. 1, 1992, pp. 23–29.

656 Asgarpoor, S.,"Generation System Reliability Evaluation with Fuzzy Data," *Proceedings of the 57th Annual American Power Conference*, Vol. 57-1, 1995, pp. 631–635.
657 Azam, M., Tu, F., Pattipati, K., "Condition-Based Predictive Maintenance of Industrial Power Systems," *Proceedings of the SPIE Conference on Components and Systems Diagnostics, Progrnostics, and Health Management II*, Vol. 4733, 2002, pp. 133–144.
658 Bagen, I., Billinton, R., "Evaluation of Different Operating Strategies in Small Stand-Alone Power Systems," *IEEE Transactions on Energy Conversion,* Vol. 20, No. 3, 2005, pp. 654–660.
659 Balijepalli, N., Venkata, S.S., Christie, R.D., "Modeling and Analysis of Distribution Reliability Indices," *IEEE Transactions on Power Delivery,* Vol. 19, No. 4, 2004, pp. 1950–1955.
660 Barinov, V.A., Volkov, G.A., Kalita, V.V., et al, "Improvement of Standards on the Functional Reliability of Electric Power Systems," *Elektrichestvo,* No. 7, 1993, pp. 1–9.
661 Bellomo, P., Donaldson, A., MacNair, D., "B-Factory Intermediate DC Magnet Power Systems Reliability Modeling and Results," *Proceedings of the IEEE Particle Accelerator Conference*, Vol. 5, 2001, pp. 3684–3686.
662 Belousenko, I.V., Golubev, S.V., Dil'man, M.D., Popyrin, L.S., "Exploration of Reliability of the Local Power Systems," *Izvestiya Akademii Nauk.Energetika,* No. 6, 2004, pp. 48–60.
663 Bennett, L., Dodenhoeft, T., "Quality and Reliability: A Customer's Perspective," *Proceedings of the IEEE Applied Power Electronics Conference and Exhibition,* 1991, pp. 120–122.
664 Bertling, L., Allan, R., Eriksson, R., "A Reliability-Centred Asset Maintenance Method for Assessing the Impact of Maintenance in Power Distribution Systems," *Proceedings of the IEEE Power Engineering Society General Meeting,* Vol. 3, 2005, pp. 2649.
665 Beshir, M.J., Farag, A.S., Cheng, T.C., "New Comprehensive Reliability Assessment Framework for Power Systems," *Energy Conversion and Management,* Vol. 40, No. 9, 1999, pp. 975–1007.
666 Bhavaraju, M.P.,"Role of a Composite System Reliability Model in Adequacy Assessment," *Proceedings of the IEEE Power Engineering Society Transmission and Distribution Conference*, Vol. 1, 2002, pp. 16–17.
667 Bie, Z., Wang, X., "Studies on Models and Algorithms of Reliability Evaluation for Cascading Faults of Complicated Power Systems," *Dianli Xitong Zidonghue/ Automation of Electric Power Systems,* Vol. 25, No. 20, 2001, pp. 30–34.
668 Billinton, R., Aboreshaid, S., "Security Evaluation of Composite Power Systems," *IEE Proceedings: Generation, Transmission and Distribution,* Vol. 142, No. 5, 1995, pp. 511–516.
669 Billinton, R., Allan, R.N., "Basic Power System Reliability Concepts," *Reliability Engineering & System Safety,* Vol. 27, No. 3, 1990, pp. 365–384.
670 Billinton, R., Cui, Y., "Reliability Evaluation of Composite Electric Power Systems Incorporating Facts," *Proceedings of the Canadian Conference on Electrical and Computer Engineering*, Vol. 1, 2002, pp. 1–6.
671 Billinton, R., Goel, L., "Overall Adequacy Assessment of an Electric Power System," *IEE Proceedings, Part C: Generation, Transmission and Distribution,* Vol. 139, No. 1, 1992, pp. 57–63.

672 Billinton, R., Hua, C., Ghajar, R., "Time-Series Models for Reliability Evaluation of Power Systems Including Wind Energy," *Microelectronics and Reliability,* Vol. 36, No. 9, 1996, pp. 1253–1261.
673 Billinton, R., Karki, R., "Maintaining Supply Reliability of Small Isolated Power Systems using Renewable Energy," *IEE Proceedings: Generation, Transmission and Distribution,* Vol. 148, No. 6, 2001, pp. 530–534.
674 Billinton, R., Karki, R., "Reliability/cost Implications of Utilizing Photovoltaics in Small Isolated Power Systems," *Reliability Engineering and System Safety,* Vol. 79, No. 1, 2003, pp. 11–16.
675 Billinton, R., Khan, E., "A Security Based Approach to Composite Power System Reliability Evaluation," *IEEE Transactions on Power Systems,* Vol. 7, No. 1, 1992, pp. 65–72.
676 Billinton, R., Khan, M.E., "Security Considerations in Composite Power System Reliability Evaluation," *Proceedings of the Third International Conference on Probabilistic Methods Applied to Electric Power Systems,* No. 338, 1991, pp. 58–63.
677 Billinton, R., Kumar, S., "Indices for use in Composite Generation and Transmission System Adequacy Evaluation," *Electrical Power & Energy Systems,* Vol. 12, No. 3, 1990, pp. 147–155.
678 Billinton, R., Li, W., "Composite System Reliability Assessment using a Monte Carlo Approach," *Proceedings of the Third International Conference on Probabilistic Methods Applied to Electric Power Systems,* No. 338, 1991, pp. 53–57.
679 Billinton, R., Oprisan, M., Clark, I.M., "Reliability Data Base for Performance and Predictive Assessment of Electric Power Systems," *Proceedings of the Third International Conference on Probabilistic Methods Applied to Electric Power Systems,* No. 338, 1991, pp. 255–260.
680 Billinton, R., Oteng-Adjei, J., "Utilization of Interrupted Energy Assessment Rates in Generation and Transmission System Planning," *IEEE Transactions on Power Systems,* Vol. 6, No. 3, 1991, pp. 1245–1253.
681 Billinton, R., Pandey, M., "Quantitative Reliability Assessment of Electric Power Systems in Developing Countries," *Proceedings of the 1996 Canadian Conference on Electrical and Computer Engineering,* Vol. 1, 1996, pp. 412–415.
682 Billinton, R., Tang, X., "Selected Considerations in Utilizing Monte Carlo Simulation in Quantitative Reliability Evaluation of Composite Power Systems," *Electric Power Systems Research,* Vol. 69, No. 2–3, 2004, pp. 205–211.
683 Billinton, R., Tollefson, G., Wacker, G., "Assessment of Electric Service Reliability Worth," *International Journal of Electrical Power and Energy System,* Vol. 15, No. 2, 1993, pp. 95–100.
684 Billinton, R., Wenyuan, L., "Consideration of Multi-State Generating Unit Models in Composite System Adequacy Assessment using Monte Carlo Simulation," *Canadian Journal of Electrical and Computer Engineering,* Vol. 17, No. 1, 1992, pp. 24–28.
685 Billinton, R., Wenyuan, L., "Hybrid Approach for Reliability Evaluation of Composite Generation and Transmission Systems using Monte-Carlo Simulation and Enumeration Technique," *IEE Proceedings, Part C: Generation, Transmission and Distribution,* Vol. 138, No. 3, 1991, pp. 233–241.
686 Billinton, R., Zhang, W. "Algorithm for Failure Frequency and Duration Assessment of Composite Power Systems," *IEE Proceedings: Generation, Transmission and Distribution,* Vol. 145, No. 2, 1998, pp. 117–122.

687 Billinton, R., Zhang, W., "Cost Related Reliability Evaluation of Bulk Power Systems," *International Journal of Electrical Power and Energy System,* Vol. 23, No. 2, 2001, pp. 99–112.
688 Billinton, R., Zhang, W., "Cost-Related Reliability Evaluation of Interconnected Bulk Power Systems using an Equivalent Approach," *Electric Machines and Power Systems,* Vol. 28, No. 7, 2000, pp. 793–810.
689 Billinton, R., Zhang, W., "Load Duration Curve Incorporation in the Reliability Evaluation of Bulk Power Systems," *Electric Power Components and Systems,* Vol. 30, No. 1, 2002, pp. 89–105.
690 Bollen, M.H.J."Reliability Analysis of Industrial Power Systems Taking into Account Voltage Sags," *Proceedings of the 1993 IEEE Industry Applications Meeting*, Vol. 2, 1993, pp. 1461–1468.
691 Bollen, M.H.J., Dirix, P.M.E., "Simple Model for Post-Fault Motor Behavior for reliability/power Quality Assessment of Industrial Power Systems," *IEE Proceedings: Generation, Transmission and Distribution,* Vol. 143, No. 1, 1996, pp. 56–60.
692 Bollen, M.H.J., Verstappen, K.F.L., Massee, P., "Method to Include Security in the Reliability Analysis of Power Systems," *Proceedings of the IASTED International Conference on Power Systems and Engineering*, 1992, pp. 106–108.
693 Bollen, M.H.J.,"Reliability Analysis of Industrial Power Systems with Sensitive Loads," *Microelectronics and Reliability,* Vol. 35, No. 9–10, 1995, pp. 1333–1345.
694 Bramley, J.S., Irving, N.B., Warty, P., "Focus on Network Reliability: Power and Infrastructure," *AT&T Technical Journal,* Vol. 73, No. 4, 1994, pp. 22–28.
695 Brandao, A.F.J., Senger, E.C., "Reliability of Digital Relays with Self-Checking Methods," *International Journal of Electrical Power and Energy System,* Vol. 15, No. 2, 1993, pp. 59–63.
696 Breipohl, A.M., Lee, F.N., "Stochastic Load Model for use in Operating Reserve Evaluation," *Proceedings of the Third International Conference on Probabilistic Methods Applied to Electric Power Systems*, No. 338, 1991, pp. 123–126.
697 Bretthauer, G., Gamaleja, T., Jacobs, A., Wilfert, H., "Determination of Favourable Maintenance Schedules for Improving the Power System Reliability," *Proceedings of the 11th Triennial World Congress of the International Federation of Automatic Control,* Vol. 6, 1991, pp. 109–114.
698 Burns, S., Gross, G.E., "Value of Service Reliability," *IEEE Transactions on Power Systems,* Vol. 5, No. 3, 1990, pp. 825–834.
699 Carr, W.,"Predictive Distribution Reliability Analysis Considering Post Fault Restoration and Coordination Failure," *Proceedings of the Rural Electronic Power Conference*, 2002, pp. 3–1.
700 Castro, R.M.G., Ferreira, L.A.F.M., "A Comparison between Chronological and Probabilistic Methods to Estimate Wind Power Capacity Credit," *IEEE Transactions on Power Systems,* Vol. 16, No. 4, 2001, pp. 904–909.
701 Chamberlin, D.M., Pidcock, D.J., "The Northeast Utilities Distribution Disturbance and Interruption Monitoring System," *IEEE Transactions on Power Delivery,* Vol. 6, No. 1, 1991, pp. 267–274.
702 Chan, T., Liu, C., Choe, J., "Implementation of Reliability-Centered Maintenance for Circuit Breakers," *Proceedings of the IEEE Power Engineering Society General Meeting*, Vol. 1, 2005, pp. 684–690.
703 Chassin, D.P., Posse, C. ,"Evaluating North American Electric Grid Reliability using the Barabasi-Albert Network Model," *Physica A: Statistical Mechanics and its Applications,* Vol. 355, No. 2–4, 2005, pp. 667–677.

704 Chen, H., Zhou, J., Tang, N., "Events Expansion Method for Composite-System Reliability Evaluation," *Zhongguo Dianji Gongcheng Xuebao/Proceedings of the Chinese Society of Electrical Engineering*, Vol. 11, No.l, 1991, pp. 84–91.
705 Chen, J., Zhao, J., Guo, Y., "Grey Relational and Fuzzy Nearness Analyses on the Reliability Study of Power Systems," *Zhongguo Dianji Gongcheng Xuebao/Proceedings of the Chinese Society of Electrical Engineering*, Vol. 22, No. 1, 2002, pp. 59–63.
706 Chen, L.N., Toyoda, J., "Maintenance Scheduling Based on Two-Level Hierarchical Structure to Equalize Incremental Risk," *IEEE Transactions on Power Systems*, Vol. 5, No. 4, 1990, pp. 1510–1516.
707 Chen, Q.,"Comparative Study of Power Generation System Reliability Including the Consideration of Energy Limited Units," *Proceedings of the Third International Conference on Probabilistic Methods Applied to Electric Power Systems*, No. 338, 1991, pp. 20–25.
708 Chen, X., Chen, G., Zhang, W., Wang, H., "Long, Medium and Short-Term Energy Management Systems of Large Power Systems," *Zhongguo Dianji Gongcheng Xuebao/Proceedings of the Chinese Society of Electrical Engineering*, Vol. 14, No. 6, 1994, pp. 41–48.
709 Chen, Y., Bose, A., "Security Analysis for Voltage Problems using a Reduced Model," *IEEE Transactions on Power Systems*, Vol. 5, No. 3, 1990, pp. 933–940.
710 Chen, Y., Ren, Z., Huang, W., "Model and Analysis of Power System Reliability Evaluation Considering Weather Change," *Dianli Xitong Zidonghua/Automation of Electric Power Systems*, Vol. 28, No. 21, 2004, pp. 17–21.
711 Chen, Y., Ren, Z., Liang, Z., Huang, W., "Study on Capacity Model for Reliability Evaluation of HVDC System," *Power System Technology*, Vol. 29, No. 10, 2005, pp. 9–13.
712 Chhalotra, G.P., Tripathi, R.N., Balakrishana, M., Rao, A.C., "Simulation of Failure Rates, Mean Time to Failure and Reliability of Power Networks Working Under Fault Conditions," *Modelling, Simulation & Control A: General Physics (Matter & Waves), Electrical & Electronics Engineering*, Vol. 27, No. 3, 1990, pp. 23–42.
713 Choi, J., Kim, H., Moon, S., Moon, Y., Billinton, R., "Nodal Probabilistic Production Cost Simulation and Reliability Evaluation at Load Points of Composite Power Systems," *Proceedings of the Universities Power Engineering Conference*, Vol. 36, 2001, pp. 1697–1702.
714 Choi, J., Thomas, R., Wang, Z., El-Keib, A.A., Billinton, R., "A Study on Probabilistic Optimal Reliability Criterion Determination in Composite Power System Expansion Planning," *Proceedings of the IEEE Power Engineering Society General Meeting*, Vol. 2, 2005, pp. 1277–1284.
715 Chowdhury, A.A., Koval, D.O., "Considerations of Relevant Factors in Setting Distribution System Reliability Standards," *Proceedings of the 2004 IEEE Power Engineering Society General Meeting*, Vol. 1, 2004, pp. 9–15.
716 Chowdhury, A.A., Koval, D.O., "Delivery Point Reliability Measurement," *Proceedings of the 1995 Annual IEEE Meeting on Industrial and Commercial Power Systems*, 1995, pp. 127–134.
717 Chowdhury, A.A., Koval, D.O., "Generation Reliability Impacts of Industry-Owned Distributed Generation Sources," *Proceedings of the IEEE Industry Applications Conference*, Vol. 2, 2003, pp. 1321–1327.

718 Chowdhury, A.A., Mielnik, T.C., Lawton, L.E., Sullivan, M.J., Katz, A., "System Reliability Worth Assessment at a Midwest Utility – Survey Results for Industrial, Commercial and Institutional Customers," *Proceedings of the International Conference on Probabilistic Methods Applied to Power Systems*, 2004, pp. 756–762.
719 Chowdhury, N., Billinton, R., "A Reliability Test System for Educational Purposes – Spinning Reserve Studies in Isolated and Interconnected Systems," *IEEE Transactions on Power Systems,* Vol. 6, No. 4, 1991, pp. 1578–1583.
720 Conejo, A.J., Garcia-Bertrand, R., Diaz-Salazar, M., "Generation Maintenance Scheduling in Restructured Power Systems," *IEEE Transactions on Power Systems,* Vol. 20, No. 2, 2005, pp. 984–992.
721 Cooper, D.,"Improve Your Card Power System's Reliability," *Electronic Design,* Vol. 53, No. 6, 2005, pp. 71–77.
722 Deng, Z., Singh, C., "A New Approach to Reliability Evaluation of Interconnected Power Systems Including Planned Outages and Frequency Calculations," *IEEE Transactions on Power Systems,* Vol. 7, No. 2, 1992, pp. 734–743.
723 Dialynas, E.N., Koskolos, N.C., "Evaluating the Reliability Performance of Industrial Power Systems with Stand-by and Emergency Generating Facilities," *Electric Power Systems Research,* Vol. 20, No. 2, 1991, pp. 143–155.
724 Dialynas, E.N., Koskolos, N.C., "Reliability Modeling and Evaluation of HVDC Power Transmission Systems," *IEEE Transactions on Power Delivery,* Vol. 9, No. 2, 1994, pp. 872–878.
725 Dialynas, E.N., Koskolos, N.C., Agoris, D., "Reliability Assessment of Autonomous Power Systems Incorporating HVDC Interconnection Links," *IEEE Transactions on Power Delivery,* Vol. 11, No. 1, 1996, pp. 519–525.
726 Dialynas, E.N., Papakammenos, D.J., Koskolos, N.C., "Integration of Non-Utility Generating Facilities into the Reliability Assessment of Composite Generation and Transmission Power Systems," *IEEE Transactions on Power Systems,* Vol. 12, No. 1, 1997, pp. 464–470.
727 Ding, M., Luo, C., "Calculation of the Reliability Indices for Operation Reserve," *Zhongguo Dianji Gongcheng Xuebao/Proceedings of the Chinese Society of Electrical Engineering,* Vol. 11, No. 3, 1991, pp. 51–58.
728 El-Serafi, A.M., Faried, S.O., "Effect of Sequential Reclosure of Multi-Phase System Faults on Turbine-Generator Shaft Torsional Torques," *IEEE Transactions on Power Systems,* Vol. 6, No. 4, 1991, pp. 1380–1388.
729 El-Zeftawy, A.A.,"Modelling and Evaluating the Reliability of a Power Generation System," *Modelling, Measurement & Control A,* Vol. 54, No. 4, 1994, pp. 1–8.
730 Farag, A.S., Al-Baiyat, S., "Optimal Design of Power Systems Under Constraints of Reliability and Cost," *Energy Conversion and Management,* Vol. 38, No. 7, 1997, pp. 637–645.
731 Fockens, S., van Wijk, A.J.M., Turkenburg, W.C., Singh, C., "A Concise Method for Calculating Expected Unserved Energy in Generating System Reliability Analysis," *IEEE Transactions on Power Systems,* Vol. 6, No. 3, 1991, pp. 1085–1091.
732 Fokin, Y.A., Aleksandrov, S.O., "A Method for Calculating Reliability of Electrical Energy Sources of Electrical Power Systems when Planning their," *Elektrichestvo,* No. 7, 1996, pp. 6–12.
733 Fong, C.C., Grigg, C.H., "Transmission Outage Models for Sub-Transmission Network Reliability Evaluation," *IEEE Transactions on Power Systems,* Vol. 6, No. 2, 1991, pp. 605–612.

734 Franklin, J.B., Parker, D., Cosby, L., "Testing Electrical Power Systems for Safety and Reliability," *IEEE Annual Textile, Fiber and Film Industry Technical Conference,* 1999, pp. 1–7.
735 Geber Melo, A.C., Pereira, M.V.F., Leite da Silva,Armando M., "Frequency and Duration Calculations in Composite Generation and Transmission Reliability Evaluation," *IEEE Transactions on Power Systems,* Vol. 7, No. 2, 1992, pp. 469–476.
736 Ginn, J.W., "Testing and Development of a 30-kVA Hybrid Inverter: Lessons Learned and Reliability Implications," *Progress in Photovoltaics: Research and Applications,* Vol. 7, No. 3, 1999, pp. 191–198.
737 Goel, L., Billinton, R., "Utilization of Interrupted Energy Assessment Rates to Evaluate Reliability Worth in Electric Power Systems," *IEEE Transactions on Power Systems,* Vol. 8, No. 3, 1993, pp. 929–936.
738 Goel, L., Wong, C.K., Wee, G.S., "Educational Software Package for Reliability Evaluation of Interconnected Power Systems," *IEEE Transactions on Power Systems,* Vol. 10, No. 3, 1995, pp. 1147–1153.
739 Graziano, R.P., Kruse, V.J., Rankin, G.L., "Systems Analysis of Protection System Coordination: A Strategic Problem for Transmission and Distribution Reliability," *IEEE Transactions on Power Delivery,* Vol. 7, No. 2, 1992, pp. 720–726.
740 Grinval'd, I.Y.,"Planning Measures to Provide Reliability and Safety of Hydraulic Structures at Hydroelectric Stations of the Kola and Karelia Power Systems," *Hydrotechnical Construction (English translation of Gidrotekhnicheskoe Stroitel'stvo),* Vol. 30, No. 3, 1996, pp. 133.
741 Gubbala, N., Singh, C., "Fast and Efficient Method for Reliability Evaluation of Interconnected Power Systems – Preferential Decomposition Method," *IEEE Transactions on Power Systems,* Vol. 9, No. 2, 1994, pp. 644–652.
742 Guo, Y., "Reliability of Power Systems and Power Equipment," *Dianli Xitong Zidonghue/Automation of Electric Power Systems,* Vol. 25, No. 17, 2001, pp. 53–56.
743 Gupta, P.P., Sharma, S., Sharma, R.K., "Two-Unit Standby Power Plant with Imperfect Switching," *Microelectronics and Reliability,* Vol. 30, No. 5, 1990, pp. 865–867.
744 Hamoud, G.,"Probabilistic Assessment of Interconnection Assistance between Power Systems," *IEEE Transactions on Power Systems,* Vol. 13, No. 2, 1998, pp. 535–540.
745 Han, Z., Qian, Y., Wen, F. ,"Tabu Search Approach to Fault Diagnosis in Power Systems using Fuzzy Abductive Inference," *Qinghua Daxue Xuebao/Journal of Tsinghua University,* Vol. 39, No. 3, 1999, pp. 56–60.
746 Hassan, S.S., Rastgoufard, P., "Detection of Power System Operation Violations Via Fuzzy Set Theory," *Electric Power Systems Research,* Vol. 38, No. 2, 1996, pp. 83–90.
747 He, J., Li, J., "Seismic Reliability Analysis of Large Electric Power Systems," *Earthquake Engineering and Engineering Vibration,* Vol. 3, No. 1, 2004, pp. 51–55.
748 Hirota, A., Kikuchi, M., Owaki, T., Tani, Y., "Dependable, Open and Real-Time Architecture for Power Systems – DORA-Power," *Hitachi Review,* Vol. 49, No. 2, 2000, pp. 48–52.
749 Hsu, Y., Lee, Y., Jien, J., et al, "Operating Reserve and Reliability Analysis of the Taiwan Power System," *IEE Proceedings, Part C: Generation, Transmission and Distribution,* Vol. 137, No. 5, 1990, pp. 349–357.

750 Karki, R., Billinton, R., "Reliability/cost Implications of PV and Wind Energy Utilization in Small Isolated Power Systems," *IEEE Transactions on Energy Conversion,* Vol. 16, No. 4, 2001, pp. 368–373.
751 Karki, R., Billinton, R., "Reliability/cost Implications of Utilizing Wind Energy in Small Isolated Power Systems," *Wind Engineering,* Vol. 24, No. 5, 2000, pp. 379–388.
752 Koval, D.O."Transmission Equipment Reliability Data from Canadian Electrical Association," *IEEE Transactions on Industry Applications,* Vol. 32, No. 6, 1996, pp. 1431–1439.
753 Koval, D.O., Ratusz, J., "Substation Reliability Simulation Model," *IEEE Transactions on Industry Applications,* Vol. 29, No. 5, 1993, pp. 1012–1017.
754 Koval, D.O.,"Zone-Branch Reliability Methodology for Analyzing Industrial Power Systems," *IEEE Transactions on Industry Applications,* Vol. 36, No. 5, 2000, pp. 1212–1218.
755 Lang, B.P., Pahwa, A., "Power Distribution System Reliability Planning using a Fuzzy Knowledge-Based Approach," *IEEE Transactions on Power Delivery,* Vol. 15, No. 1, 2000, pp. 279–284.
756 Lee, F.N., Lin, M., Breipohl, A.M., "Evaluation of the Variance of Production Cost using a Stochastic Outage Capacity State Model," *IEEE Transactions on Power Systems,* Vol. 5, No. 4, 1990, pp. 1061–1067.
757 Lee, H., Lee, H. "Method for Assessing the Electric Power System Reliability of Multiple-Engined Aircraft," *Journal of Aircraft,* Vol. 30, No. 3, 1993, pp. 413–414.
758 Li, J., He, J., Li, T., "Seismic Reliability Analysis of Large Electric Power Systems," *Harbin Jianzhu Daxue Xuebao/Journal of Harbin University of Civil Engineering and Architecture,* Vol. 35, No. 1, 2002, pp. 7–11.
759 Li, Z., Li, W., Liu, Y., "Fault-Traverse Algorithm of Radial-Distribution-System-Reliability Evaluation," *Dianli Xitong Zidonghue/Automation of Electric Power Systems,* Vol. 26, No. 2, 2002, pp. 53–56.
760 Liu, B., Xie, K., Ma, C., Xu, D., Zhou, J., Zhou, N., "Section Algorithm of Reliability Evaluation for Complex Medium Voltage Electrical Distribution Networks," *Zhongguo Dianji Gongcheng Xuebao/Proceedings of the Chinese Society of Electrical Engineering,* Vol. 25, No. 4, 2005, pp. 40–45.
761 Liu, B., Xie, K., Zhang, H., Zhou, Y., "Reliability Analysis of Typical Connection Modes in HV Distribution Network," *Power System Technology,* Vol. 29, No. 14, 2005, pp. 45–48.
762 Liu, H., Cheng, L., Sun, Y., Zheng, W., "Reliability Evaluation of Hybrid AC/DC Power Systems," *Power System Technology,* Vol. 28, No. 23, 2004, pp. 27–31.
763 Loparo, K.A., Abdel-Malek, F., "Probabilistic Approach to Dynamic Power System Security," *IEEE Transactions on Circuits and Systems,* Vol. 37, No. 6, 1990, pp. 787–798.
764 Love, D.J."Reliability of Utility Supply Configurations for Industrial Power Systems," *IEEE Transactions on Industry Applications,* Vol. 30, No. 5, 1994, pp. 1303–1308.
765 Maghraby, H.A.M., Allan, R.N., "Application of DC Equivalents to the Reliability Evaluation of Composite Power Systems," *IEEE Transactions on Power Systems,* Vol. 14, No. 1, 1999, pp. 355–361.
766 Makinen, A., Partanen, J., Lakervi, E. "Practical Approach for Estimating Future Outage Costs in Power Distribution Networks," *IEEE Transactions on Power Delivery,* Vol. 5, No. 1, 1990, pp. 311–316.

767 Massim, Y., Zeblah, A., Meziane, R., Benguediab, M., Ghouraf, A., "Optimal Design and Reliability Evaluation of Multi-State Series-Parallel Power Systems," *Nonlinear Dynamics,* Vol. 40, No. 4, 2005, pp. 309–321.
768 Matijevics, I., Jozsa, L., "Expert-System-Assisted Reliability Analysis of Electric Power Networks," *Engineering Applications of Artificial Intelligence,* Vol. 8, No. 4, 1995, pp. 449–460.
769 Medicherla, T.K.P., Chau, M., Zigmund, R.E., Chen, K., "Transmission Station Reliability Evaluation," *IEEE Transactions on Power Systems,* Vol. 9, No. 1, 1994, pp. 295–304.
770 Mili, L., Qiu, Q., Phadke, A.G., "Risk Assessment of Catastrophic Failures in Electric Power Systems," *International Journal of Critical Infrastructures,* Vol. 1, No. 1, 2004, pp. 38–63.
771 Miller, J.M., Emadi, A., Rajarathnam, A.V., Ehsani, M., "Current Status and Future Trends in More Electric Car Power Systems," *IEEE Vehicular Technology Conference,* Vol. 2, 1999, pp. 1380–1384.
772 Misra, R.B., Patra, S. "New Method for Reliability Evaluation of Composite Power Systems," *Microelectronics and Reliability,* Vol. 34, No. 6, 1994, pp. 983–998.
773 Mitra, J., Singh, C., "Pruning and Simulation for Determination of Frequency and Duration Indices of Composite Power Systems," *IEEE Transactions on Power Systems,* Vol. 14, No. 3, 1999, pp. 899–905.
774 Modarres, M., Farrokhzad, D., "Reliability Consideration in Optimization of Cascaded Hydrothermal Power Systems," *International Journal of Power and Energy Systems,* Vol. 23, No. 1, 2003, pp. 6–14.
775 Moro, L.M., Ramos, A., "Goal Programming Approach to Maintenance Scheduling of Generating Units in Large Scale Power Systems," *IEEE Transactions on Power Systems,* Vol. 14, No. 3, 1999, pp. 1021–1027.
776 Mottate, Y., Saitoh, H., Toyoda, J., "Composite Power Systems Reliability Evaluation Based on Demand-Side Reserve," *Electrical Engineering in Japan (English translation of Denki Gakkai Ronbunshi),* Vol. 114, No. 7, 1994, pp. 68–78.
777 Nahman, J.M., Graovac, M.B., "A Method for Evaluating the Frequency of Deficiency-States of Electric-Power Systems," *IEEE Transactions on Reliability,* Vol. 39, No. 3, 1990, pp. 265–272.
778 Newby, R.A., Lippert, T.E., Alvin, M.A., Burck, G.J., Sanjana, Z.N., "Status of Westinghouse Hot Gas Filters for Coal and Biomass Power Systems," *Journal of Engineering for Gas Turbines and Power, Transactions of the ASME,* Vol. 121, No. 3, 1999, pp. 401–408.
779 Ochoa, J.R. and Garrison, D.L. "Application of Reliability Criteria in Power Systems Planning," Vol. 56, No. 2, 1994, pp. 1114–1120.
780 O'Donnell, P., Braun, W.F., Heising, C.R., Khera, P.P., Kornblit, M., McDonald, K.D., "Survey Results of Low-Voltage Circuit Breakers as found during Maintenance Testing: Working Group Report," *IEEE Transactions on Industry Applications,* Vol. 33, No. 5, 1997, pp. 1367–1369.
781 Okuda, K., Watanabe, H., Yamazaki, K., Baba, T., "Fault Restoration Operation Scheme in Secondary Power Systems using Case-Based Reasoning," *Electrical Engineering in Japan (English translation of Denki Gakkai Ronbunshi),* Vol. 110, No. 2, 1990, pp. 47–59.

782 Oteng-Adjei, J., Billinton, R., "Evaluation of Interrupted Energy Assessment Rates in Composite Systems," *IEEE Transactions on Power Systems,* Vol. 5, No. 4, 1990, pp. 1317–1323.
783 Papakammenos, D.J., Dialynas, E.N., "Reliability and Cost Assessment of Power Transmission Networks in the Competitive Electrical Energy Market," *IEEE Transactions on Power Systems,* Vol. 19, No. 1, 2004, pp. 390–398.
784 Pereira, M.V.F., Pinto, L.M.V.G., "A New Computational Tool for Composite Reliability Evaluation," *IEEE Transactions on Power Systems,* Vol. 7, No. 1, 1992, pp. 258–264.
785 Pitz, V., Weber, T., "Forecasting of Circuit-Breaker Behaviour in High-Voltage Electrical Power Systems: Necessity for Future Maintenance Management," *Journal of Intelligent and Robotic Systems: Theory and Applications,* Vol. 31, No. 1–3, 2001, pp. 223–228.
786 Preston, E.G., Grady, W.M. "Efficient Method for Calculating Power System Production Cost and Reliability," *IEE Proceedings, Part C: Generation, Transmission and Distribution,* Vol. 138, No. 3, 1991, pp. 221–227.
787 Qu, Z., Xie, G., Zhao, X., "Development of Service Oriented Architecture of Supervisory Information System for Power Plant Generating Equipment Reliability," *Power System Technology,* Vol. 28, No. 20, 2004, pp. 23–27.
788 Ren, Z., Chen, J., Huang, W., Li, B., "Model and Algorithm of Reliability Evaluation for Large Power Systems," *Dianli Xitong Zidonghue/Automation of Electric Power Systems,* Vol. 23, No. 5, 1999, pp. 4–10.
789 Ren, Z., Wan, G., Huang, J., Huang, W., Gao, Z., "Prediction of Original Reliability Parameter of Power Systems by an Improved Grey Model," *Dianli Xitong Zidonghua/Automation of Electric Power Systems,* Vol. 27, No. 4, 2003, pp. 37–40.
790 Roda, V.O., Trindade, O.J., "On the Effect of Power-Line Disturbances on Microcomputer Performance," *Microelectronics and Reliability,* Vol. 31, No. 2–3, 1991, pp. 229–235.
791 Romero-Romero, D., Gomez-Hernandez, J.A., Robles-Garcia, J., "Reliability Optimisation of Bulk Power Systems Including Voltage Stability," *IEE Proceedings: Generation, Transmission and Distribution,* Vol. 150, No. 5, 2003, pp. 561–566.
792 Sallam, A.A., Desouky, M., Desouky, H., "Shunt Capacitor Effect on Electrical Distribution System Reliability," *IEEE Transactions on Reliability,* Vol. 43, No. 1, 1994, pp. 170–176.
793 Sanghvi, A.P., Balu, N.J., Lauby, M.G., "Power System Reliability Planning Practices in North America," *IEEE Transactions on Power Systems,* Vol. 6, No. 4, 1991, pp. 1485–1492.
794 Sanghvi, A.P.,"Measurement and Application of Customer Interruption costs/value of Service for Cost-Benefit Reliability Evaluation: Some Commonly Raised Issues," *IEEE Transactions on Power Systems,* Vol. 5, No. 4, 1990, pp. 1333–1344.
795 Schilling, M.T., Do Coutto Filho, M.B., "Power Systems Operations Reliability Assessment in Brazil," *Quality and Reliability Engineering International,* Vol. 14, No. 3, 1998, pp. 153–158.
796 Schilling, M.T., Souza, J.C.S., Alves da Silva, A.P., Do Coutto Filho, M.B., "Power Systems Reliability Evaluation using Neural Networks," *International Journal of Engineering Intelligent Systems for Electrical Engineering and Communications,* Vol. 9, No. 4, 2001, pp. 219–226.

797 Shahidepour, M., Allan, R., Anderson, P., et al, "Effect of Protection Systems on Bulk Power Reliability Evaluation," *IEEE Transactions on Power Systems,* Vol. 9, No. 1, 1994, pp. 198–205.
798 Silva, E.L., Mesa, S.E.C., Morozowski, M. ,"Transmission Access Pricing to Wheeling Transactions: A Reliability Based Approach," *IEEE Transactions on Power Systems,* Vol. 13, No. 4, 1998, pp. 1481–1486.
799 Simpson, B.,"Progress Towards Integration of Reliability Concepts in a Segregated Electric Utility System," *Reliability Engineering & System Safety,* Vol. 46, No. 1, 1994, pp. 109–112.
800 Singh, C., Chen, Q., "Modeling of Energy Limited Units in the Reliability Evaluation of Multi-Area Electrical Power Systems," *IEEE Transactions on Power Systems,* Vol. 5, No. 4, 1990, pp. 1364–1373.
801 Singh, C., Chintaluri, G.M., "Reliability Evaluation of Interconnected Power Systems using a Multi-Parameter Gamma Distribution," *International Journal of Electrical Power and Energy System,* Vol. 17, No. 2, 1995, pp. 151–160.
802 Singh, C., Deng, Z., "New Algorithm for Multi-Area Reliability Evaluation. Simultaneous Decomposition-Simulation Approach," *Electric Power Systems Research,* Vol. 21, No. 2, 1991, pp. 129–136.
803 Singh, C., Gubbala, N., "Reliability Evaluation of Interconnected Power Systems Including Jointly Owned Generators," *IEEE Transactions on Power Systems,* Vol. 9, No. 1, 1994, pp. 404–412.
804 Singh, C., Mitra, J., "Reliability Analysis of Emergency and Standby Power Systems," *IEEE Industry Applications Magazine,* Vol. 3, No. 5, 1997, pp. 41–47.
805 Singh, C., Patton, A.D., Kumar, M., Wagner, H.A., "A Simulation Model for Reliability Evaluation of Space Station Power Systems," *IEEE Transactions on Industry Applications,* Vol. 27, No. 2, 1991, pp. 331–334.
806 Skuletic, S., Allan, R.N., "Reliability Analysis – a Mean for Improvement of Power System Characteristics," *Proceedings of the Universities Power Engineering Conference,* Vol. 1, 1999, pp. 38–41.
807 Song, X., Tan, Z., "Application of Improved Importance Sampling Method in Power System Reliability Evaluation," *Power System Technology,* Vol. 29, No. 13, 2005, pp. 56–59.
808 Song, Y., Zhou, S., Lu, Z., Zhang, R., "A New Calculation Method for Optimal Reliability Indices of Composite Power System using GA," *Power System Technology,* Vol. 28, No. 15, 2004, pp. 25–30.
809 Stillman, R.H.,"Liability of Poor Reliability: Legal Ramifications when Power Systems Fail," *IEEE Power and Energy Magazine,* Vol. 2, No. 4, 2004, pp. 88–86.
810 Su, C., Lii, G., "Reliability Planning for Composite Electric Power Systems," *Electric Power Systems Research,* Vol. 51, No. 1, 1999, pp. 23–31.
811 Sveshnikov, V.I., Fokin, Y.A., "Calculation Methods and Hardware Providing for Reliability of Electric Power Systems in Dynamic Conditions," *Applied Energy Research: Russian Journal of Fuel, Power, and Heat Systems,* Vol. 33, No. 5, 1995, pp. 77–81.
812 Tan, L., Wu, Z., "Research of a Three-Phase Short Current Limiter in Distribution Power Systems," *Zhejiang Daxue Xuebao (Gongxue Ban)/Journal of Zhejiang University (Engineering Science Edition,* Vol. 36, No. 1, 2002, pp. 101–104.
813 Tan, Z.,"On Reliability Optimization for Power Generation Systems," *Journal of Systems Engineering and Electronics,* Vol. 16, No. 3, 2005, pp. 697–703.

814 Tanrioven, M., Wu, Q.H., Turner, D.R., Kocatepe, C., Wang, J., "A New Approach to Real-Time Reliability Analysis of Transmission System using Fuzzy Markov Model," *International Journal of Electrical Power and Energy System,* Vol. 26, No. 10, 2004, pp. 821–832.
815 Tracey, T.,"Improvements in the Cost Effectiveness and Reliability of Central Receiver Solar Power Systems," *ASME-JSES-JSME International Solar Energy Conference,* 1991, pp. 267–274.
816 Ubeda, J.R., Allan, R.N. "Reliability Assessment of Hydro-Thermal Composite Systems by Means of Stochastic Simulation Techniques," *Reliability Engineering & System Safety,* Vol. 46, No. 1, 1994, pp. 33–47.
817 Ubeda, J.R., Allan, R.N., "Sequential Simulation Applied to Composite System Reliability Evaluation," *IEE Proceedings, Part C: Generation, Transmission and Distribution,* Vol. 139, No. 2, 1992, pp. 81–86.
818 Van Casteren, J.F.L., Bollen, M.H.J., Schmieg, M.E., "Reliability Assessment in Electrical Power Systems: The Weibull-Markov Stochastic Model," *IEEE Transactions on Industry Applications,* Vol. 36, No. 3, 2000, pp. 911–915.
819 Vanzi, I.,"Seismic Reliability of Electric Power Networks: Methodology and Application," *Structural Safety,* Vol. 18, No. 4, 1996, pp. 311–327. .
820 Wang, C., Xie, Y., "A New Bayesian Network Model for Distribution System Reliability Evaluation Based on Dual Isomorphic Bayesian Network Model," *Power System Technology,* Vol. 29, No. 7, 2005, pp. 41–46.
821 Wang, G., Ding, M., Li, X., Liao, Z., "Reliability Analysis of UPS for Power System," *Dianli Xitong Zidonghua/Automation of Electric Power Systems,* Vol. 29, No. 3, 2005, pp. 40–44.
822 Wang, P., Billinton, R. "Reliability Benefit Analysis of Adding WTG to a Distribution System," *IEEE Transactions on Energy Conversion,* Vol. 16, No. 2, 2001, pp. 134–139.
823 Wang, P., Ding, Y., Xiao, Y., "Technique to Evaluate Nodal Reliability Indices and Nodal Prices of Restructured Power Systems," *IEE Proceedings: Generation, Transmission and Distribution,* Vol. 152, No. 3, 2005, pp. 390–396.
824 Wang, P., Goel, L., Ding, Y. ,"The Impact of Random Failures on Nodal Price and Nodal Reliability in Restructured Power Systems," *Electric Power Systems Research,* Vol. 71, No. 2, 2004, pp. 129–134.
825 Wang, S., Zhou, J., "Reliability Evaluation Model for Two Transmission Lines in Parallel," *Zhongguo Dianji Gongcheng Xuebao/Proceedings of the Chinese Society of Electrical Engineering,* Vol. 23, No. 9, 2003, pp. 53–56.
826 Wang, W., Loman, J.M., Arno, R.G., Vassiliou, P., Furlong, E.R., Ogden, D., "Reliability Block Diagram Simulation Techniques Applied to the IEEE Std. 493 Standard Network," *IEEE Transactions on Industry Applications,* Vol. 40, No. 3, 2004, pp. 887–895.
827 Wu, K., Wu, Z. ,"Reliability Evaluation Algorithm of Electrical Power Systems using Sensitivity Analysis," *Zhongguo Dianji Gongcheng Xuebao/Proceedings of the Chinese Society of Electrical Engineering,* Vol. 23, No. 4, 2003, pp. 53–56.
828 Wu, K., Wang, S., Zhang, A., Zhou, J., "Study on Reliability Assessment of Electrical Power Systems using RBF Neural Network," *Zhongguo Dianji Gongcheng Xuebao/Proceedings of the Chinese Society of Electrical Engineering,* Vol. 20, No. 6, 2000, pp. 9–12.

829 Wu, M., Ray, A., "Damage-Mitigating Control of Power Systems for Structural Durability and High Performance," *Journal of Engineering for Gas Turbines and Power, Transactions of the ASME,* Vol. 117, No. 2, 1995, pp. 307–313.
830 Xie, Z., Vittal, V., Centeno, V., Manimaran, G., Phadke, A.G., "An Information Architecture for Future Power Systems and its Reliability Analysis," *IEEE Transactions on Power Systems,* Vol. 17, No. 3, 2002, pp. 857–863.
831 Yu, D.C., Nguyen, T.C., Haddawy, P., "Bayesian Network Model for Reliability Assessment of Power Systems," *IEEE Transactions on Power Systems,* Vol. 14, No. 2, 1999, pp. 426–432.
832 Zhang, J., Dobson, I., Alvarado, F.L., "Quantifying Transmission Reliability Margin," *International Journal of Electrical Power and Energy System,* Vol. 26, No. 9, 2004, pp. 697–702.
833 Zhang, P., Wang, S., "Novel Interval Methods in Power System Reliability Economics," *Zhongguo Dianji Gongcheng Xuebao/Proceedings of the Chinese Society of Electrical Engineering,* Vol. 24, No. 2, 2004, pp. 71–77.
834 Zhang, Y.,"Reliability Analysis of an (N+1)-Unit Standby System with 'Preemptive Priority' Rule," *Microelectronics and Reliability,* Vol. 36, No. 1, 1996, pp. 19–26.
835 Zhang, Z., Xu, G., Yang, X., Zhang, Y., "Reliability Mesh Algorithm for Two Interconnected Power Systems with Correlated Loads," *Zhongguo Dianji Gongcheng Xuebao/Proceedings of the Chinese Society of Electrical Engineering,* Vol. 14, No. 2, 1994, pp. 54–59.
836 Zhong, B., Zhou, J., Zhao, Y., "Research on Soft Computing Models for Reliability Assessment of Bulk Power Systems," *Diangong Jishu Xuebao/Transactions of China Electrotechnical Society,* Vol. 20, No. 6, 2005, pp. 46–51.

A.2.8 Medical Equipment Reliability

837 Alexander, K., Clarkson, P.J., "Good design practice for medical devices and equipment, Part I: A review of current literature," *Journal of Medical Engineering and Technology,* Vol. 24, No. 1, 2000, pp. 5–13.
838 De Lemos, Z., "FMEA software program for managing preventive maintenance of medical equipment," *Proceedings of the IEEE 30th Annual Northeast Bioengineering Conference,* , 2004, pp. 247–248.
839 Dhillon, B.S., Methods for Performing Human Reliability and Error Analysis in Health Care, *International Journal of Health Quality Assurance*, Vol. 16, No. 6, 2003, pp.306–317.
840 Dhillon, B.S., Rajendran, M., "Human error in health care systems: Bibliography," *International Journal of Reliability, Quality and Safety Engineering,* Vol. 10, No. 1, 2003, pp. 99–117.
841 Dhillon, B.S., Reliability Technology for Manufacturers: Engineering Better Devices, *Medical Device and Diagnostic Industry (MDDI)*, Vol.23, No.10, 2001, pp.94–99.
842 Edlich, R.F., Wind, T.C., Heather, C.L., "Reliability and performance of innovative surgical double-glove hole puncture indication systems," *Journal of Long-Term Effects of Medical Implants,* Vol. 13, No. 2, 2003, pp. 69–83.

843 Garcia, M.A., Villar, R.S.G., Cardoso, A.L.R., "A Low-Cost and High-Reliability Communication Protocol for Remote Monitoring Devices," *IEEE Transactions on Instrumentation and Measurement,* Vol. 53, No. 2, 2004, pp. 612–618.
844 Garner, R.P., Mandella Jr., J.G., "Reliability of the gas supply in the Air Force Emergency Passenger Oxygen System (EPOS)," Proceedings of the *SAFE Association 42nd Annual Symposium,* 2004, pp. 5–10.
845 Jiang, H., Chen, W.R., Wang, G., "Localization error analysis for stereo X-ray image guidance with probability method," *Medical Engineering and Physics,* Vol. 23, No. 8, 2001, pp. 573–581.
846 Lo, M., "Benchmarking biomedical equipment maintenance in Hospital Authority (HA)," Proceedings of the *3rd Seminar on Appropriate Medical technology for Developing Countries,* London, 2004, pp. 55–59.
847 Lo, M., "Formulating equipment maintenance strategy in Hospital Authority (HA)," Proceedings of the *3rd Seminar on Appropriate Medical technology for Developing Countries,* London, 2004, pp. 61–64.
848 Nojima, T., Tarusawa, Y., "A new EMI test method for electronic medical devices exposed to mobile radio wave," *Electronics and Communications in Japan, Part I: Communications (English Translation of Denshi Tsushin Gakkai Ronbunshi),* Vol. 85, No. 4, 2002, pp. 1–9.
849 Phee, L., Xiao, D., Yuen, J., "Control and safety aspects of medical robots for treatment of diseases of the prostate," *Proceedings of the Institution of Mechanical Engineers.Part I: Journal of Systems and Control Engineering,* Vol. 217, No. 3, 2003, pp. 155–167.
850 Ridgway, M., "Analyzing planned maintenance (PM) inspection data by Failure Mode and Effect Analysis methodology," *Biomedical Instrumentation and Technology,* Vol. 37, No. 3, 2003, pp. 167–179.
851 Rovetta, A., "Telerobotic surgery control and safety," Proceedings of the *IEEE International Conference on Robotics and Automation,* 2000, pp. 2895–2900.
852 Sitaraman, S.K., Raghunathan, R., Hanna, C.E., "Development of virtual reliability methodology for area-array devices used in implantable and automotive applications," *IEEE Transactions on Components and Packaging Technologies,* Vol. 23, No. 3, 2000, pp. 452–461.
853 Tchoudovski, I., Schlindwein, M., and Jager, M., "Vergleichende untersuchungen zur zuverlassigkeit automatisierter externer defibrillatoren; Comparative reliability analysis of automatic external defibrillators," *Biomedizinische Technik,* Vol. 49, No. 6, 2004, pp. 153–156.
854 Thornton, K.E.B., Lazzarini, A., "The use of reliability techniques to predict the recovery rate of recovering alcohol addicts," *Journal of the American Society for Information Science and Technology,* Vol. 56, No. 4, 2005, pp. 356–363.
855 Vandenberghe, H.E.E., Van Casteren, V., Jonckheer, P., "Collecting information on the quality of prescribing in primary care using semi-automatic data extraction from GPs' electronic medical records," *International Journal of Medical Informatics,* Vol. 74, No. 5, 2005, pp. 367–376.
856 Wu, J., Zhang, R.R., Radons, S., "Vibration analysis of medical devices with a calibrated FEA model," *Computers and Structures,* Vol. 80, No. 12, 2002, pp. 1081–1086.
857 Wu, J., Zhang, R.R., Wu, Q., "Environmental vibration assessment and its applications in accelerated tests for medical devices," *Journal of Sound and Vibration,* Vol. 267, No. 2, 2003, pp. 371–383.

Author Biography

Dr. B.S. Dhillon is a professor of Engineering Management in the Department of Mechanical Engineering at the University of Ottawa. He has served as a Chairman/Director of Mechanical Engineering Department/Engineering Management Program for over ten years at the same institution. He has published over 330 articles on engineering management, reliability, safety, etc. He is or has been on the editorial boards of nine international scientific journals. In addition, Dr. Dhillon has written 30 books on various aspects of engineering management, design, reliability, safety, and quality (published by Wiley (1981), Van Nostrand (1982), Butterworth (1983), Marcel Dekker (1984), Pergamon (1986), etc). His books are being used in over 70 countries, and many of them are translated into languages such as German, Russian and Chinese. He has served as General Chairman of two international conferences on reliability and quality control held in Los Angeles and Paris in 1987.

Prof. Dhillon has served as a consultant to various organizations and bodies and has many years of experience in the industrial sector. At the University of Ottawa, he has been teaching reliability, quality, engineering management, design, and related areas for over 26 years. He has also lectured in over 50 countries, including keynote addresses at various international scientific conferences held in North America, Europe, Asia, and Africa. In March 2004, Dr. Dhillon was a distinguished speaker at the Conf./Workshop on Surgical Errors (sponsored by White House Health and Safety Committee and the Pentagon), held at the Capitol Hill (One Constitution Avenue, Washington, D.C.).

Professor Dhillon attended the University of Wales, where he received a BS in electrical and electronic engineering and an MS in mechanical engineering. He received a Ph.D. in industrial engineering from the University of Windsor.

Index

Printed in the United States
70532LV00002B/205-267